The .58- and .50-Caliber Rifles & Carbines of the Springfield Armory, 1865–1872

by

Richard A. Hosmer

North Cape Publications®, Inc.

In loving memory of Dale Jon Hosmer 1936–1997

Also, with love and appreciation to my wife, Kay Millard Hosmer: a fellow collector, and fine shot, who provided love and encouragement when I needed it the most, and who would not let me quit this project!

Many friends and/or fellow collectors have provided valuable insights, or other assistance, along the way. I would particularly like to acknowledge the help of the following: Lou Behling, Cy Bird, Richard Branum, Clifford Bunds, Graham Burnside, Jerry Chastain, James Curlovic, Dave Dodds, Art Domingos, R. Stephen Dorsey, Ted Dowd, Dušan Farrington, Sterling Fenn, Steve Fentress, Tom Fleming, Al Frasca, Dave Fullerton, J. Edward Green, Jeff Hayes, Vance Haynes, Bob Hill, Herbert Houze, Dimitri Ilyin, Burton Kellerstedt, Don Kenyon, Peter & Ray Kramer, Jack Lewis Jr., Mitch Luksich, Frank Mallory, Roy Marcot, John McCabe, Ken McPheeters, William Mook, Tom Murray, Merle Olmsted, Al Perry, John Ritzenthaler, Bob Ruebman, William Rutter, Carl Schoeppl, James Shaffer, Robert Snitselaar, Herb Spalla, Tom Trevor, Jack Trotter, Frank Tuculet, William Urick, and Robert Voliva.

I especially wish to thank Ed Hull and Craig Riesch, who generously gave of their time to edit the entire manuscript and offer useful and constructive criticism. Any errors of fact or interpretation that remain are the full responsibility of the author.

ISBN-10 1-882391-38-1 ISBN-13 978-1-882391-38-7
North Cape Publications®, Inc., P.O. Box 1027, Tustin, California 92781
800 745-9714, Fax 714 832-5302
E-mail: ncape@ix.netcom.com. Website http://www.northcapepubs.com
Printed by Delta Printing Solutions, Valencia, CA 91355

Table of Contents

Part II: Non-U.S. Ordnance Department Designs
Fabricated or Assembled at the Springfield Armory 1865 to 1871

Personalities Involved in the Transition from the Percussion Rifle Musket to the Breechloading Rifle

1865–1872

Erskine S. Allin, Master Armorer, Springfield Armory, 1865–1879. Photo courtesy of the Springfield Armory National Historic Site.

Brigadier General Alexander Byrdie Dyer, U.S. Army, Chief of Ordnance, 1864–1874. Photo courtesy of the National Archives.

Colonel Stephen V. Benét, U.S. Army, Commanding Officer, Frankford Arsenal, Chief of Ordnance, 1874–1891. Photo courtesy of the National Archives.

Major General John McAllister Schofield, U.S. Army, President, St. Louis Ordnance Board on Tactics, Small-Arms and Accoutrements, 1869, and later Commander, U.S. Army, 1888–1895. Photo courtesy of the National Archives.

Brigadier General Alfred H. Terry, U.S. Army, President, Board on Breech Loading Small-Arms and Calibre, 1872. Photo courtesy of the National Archives.

Introduction

The purpose of this book, *The .58- and .50-Caliber Springfield Rifles and Carbines of the Springfield Armory, 1865–1872*, in the "For Collectors Only®" series, is to provide a concise and accurate description of the military arms designed, developed and/or produced, under the auspices of the U.S. Army's Ordnance Department at the Springfield Armory, Massachusetts. The development of these rifles and carbines led directly to the adoption of the U.S. Model .45-70 Springfield rifle and carbine that would serve the U.S. Army in the two decades from 1873 to 1893, and even longer with the militias and National Guards of the various states.

Included in this volume are the four .58- and .50-caliber rifles adopted for general issue, the four cadet rifles made for use at the nation's military academies and at high schools and colleges which provided courses in military training, and the twelve rifles and carbines subjected to trials which were intended to find the design best suited to the "next-generation" military rifle.

The development and use of each arm will be described in the following pages. A brief historical sketch places the weaponry in the context of the times. The identifying features of each as well as markings are included. Because many of the parts of these weapons were common to one another, the usual part-by-part treatment found in the "For Collectors Only" series would be superfluous. Instead, the concentration is on noncommon parts and how they differ. Any part not specifically mentioned should be presumed to be exactly the same, including finish, as those on the standard arm. Arms from non-Ordnance Department sources (Remington and Ward-Burton) or adaptations of the issue military weapons (Sharps and Spencer) are described in greater detail.

While the author has directly examined several hundreds of the .45-, .50-, and .58-caliber Springfields of all types during thirty-five years of collecting, it must be realized that many of the guns discussed herein are quite scarce, if not rare, and thus are difficult to obtain for study. Still, most of the arms described represent direct physical examination of at least one specimen, which, in most cases, is from the author's personal collection. Outright speculation has been kept to a minimum; if something is not known to be true, it is so qualified.

Data on the types of finish for most visible components and the methods used at the Springfield Armory are described in Appendix B.

Illustrations are referenced in the text with a notation such as see Figure 1-1 or refer to Figure 2-3. The first number gives the chapter where the illustration may be found; the second number indicates its order within that chapter. All photographs are from the author's collection, unless credited otherwise.

It is acknowledged that perfectly legitimate, Springfield-produced arms are known with some feature(s) differing slightly from the "standard" versions of the limited production guns discussed here; some of these special arms had their "one-off" prototypes! Details of such "pattern" guns are really beyond the scope of this book; but, where the author has knowledge of such an item, it is mentioned.

A word about "quality": Springfield Armory did nothing crudely. Do not let anyone tell you otherwise when contemplating the purchase of a supposedly "rare" variation. Even though some very limited production arms are addressed in this book, they are still fitted and finished every bit as well as the production guns. There may be evidence of parts having been adapted from something else, but there will be no rough edges, or gaps, or misalignments, or evidence of force-fitting. If you encounter such, the odds are very good that you are looking at some form of aftermarket manipulation.

The author and publisher welcome, in fact earnestly solicit, input from fellow collectors having any corrections or further information regarding these arms, which may be incorporated into future editions of this work.

Conventions

Dimensions are provided throughout the work to allow a positive identification, yet not contribute to fakery. Some dimensions may vary slightly from those given, due to manufacturing tolerances, wood shrinkage, and years of honest wear. Small (decimal) measurements were taken with a micrometer; others may be expressed in less precise terms. Where an overall stock length measurement is given, it will include butt plate and tip (if applicable) unless specifically noted otherwise. Measurements quoted for two-piece stocks will be taken from approximate centers of end surfaces, or along bottom edge from stock toe to the rear edge of the receiver, as noted. Barrel lengths were obtained by measuring the inside of the bore from the breech face to the muzzle.

Following the practice at the time, measurements are given in inches, i.e., 1 inch or 1-3/4 inches. Measurements under 1 inch are given in decimal inches, i.e., 0.125 inch. Both standard and decimal inches may be used in some sections for clarity.

All directions are given from the shooter's point of view, i.e., looking toward the muzzle. "Right side" or "left side" refers to the side of the firearm to the shooter's right or left when the rifle is shouldered properly.

Screw-hole measurements are from the center of the hole, unless otherwise stated.

Quotation marks are used in the text to indicate factory or Ordnance Department applied markings. The quotation marks were not part of the original marking unless otherwise stated.

Serial numbers are shown divided by commas to make them more readable. Commas were *not* used in the actual markings on rifles or carbines. Not all small arms described in this text were identified by serial numbers.

Spellings used in related documents are as originally published.

Part I

U.S. Ordnance Department Rifle Designs, Manufactured at the Springfield Armory

1865 to 1872

Chapter 1: The U.S. Model 1865 .58 Caliber Springfield Rifle (aka "First Allin Alteration")

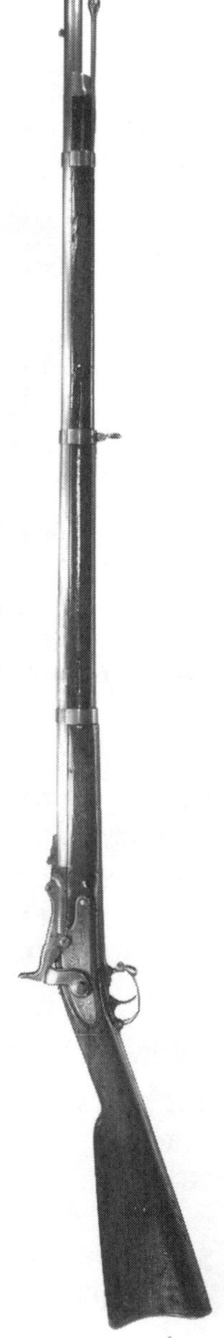

Historical Background

By the end of the Civil War, it was clear to all but the most stubborn traditionalists that general issuance of a standard breechloading arm of some sort was definitely to be desired. While breechloading arms had been around for a very long time, they all suffered one major problem: that of properly sealing the breech at the instant of discharge. With the advent of practical self-contained metallic cartridges, as typified during the war by those of the Henry and Spencer, the stage was finally set for full-scale implementation of such a design.

Erskine S. Allin, master armorer at Springfield, devised a method for converting existing rifle muskets, of which a very great number were available as war surplus, into breechloaders, see Figure 1-1. This was accomplished by cutting away a part of the upper rear end of the barrel, just ahead of the breech screw and inserting a steel block, hinged at the front, and operated by a side-mounted thumb latch, that fitted neatly into the recess in the lock plate where the drum and nipple of the percussion musket had been, see Figure 1-2. The slightly modified hammer struck a long firing pin which slanted downward and inward through the block, so as to hit the rim of the cartridge at top center.

The cartridge used was a short, squat, copper-cased rimfire, in .58 caliber, which required no change to

Fig. 1-1. The U.S. Model 1865 .58 Caliber Springfield Rifle, also known to collectors as the "First Allin Alteration."

the musket bore, or rifling, other than the reaming of a chamber and throat, see Figure 1-3. The powder charge was 60 grains, and the bullet weighed 500 grains. A specimen in the author's collection measures as follows: rim, 0.702 inch in diameter by 0.062 inch thick; case, 0.629 inch at head, 0.618 inch at mouth, 1.180 inches long. Overall length is 1.690 inches. The full designation is .58-60-500.

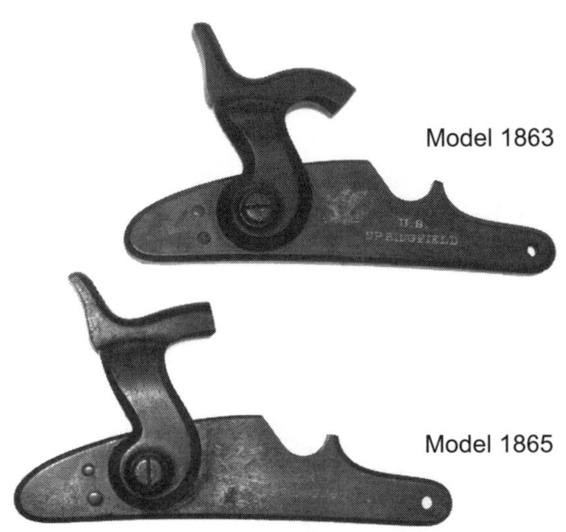

Fig. 1-2. Right-side view of the Model 1865 lock showing the lock plate and hammer compared to the 1863 percussion rifle musket lock plate and hammer.

Quantity Produced

A total of 5,000 U.S. Model 1865 .58 Caliber Rifles were manufactured, all during Fiscal Year 1866. Despite the model year designation, this arm did not see Civil War service. The U.S. Army was never fully equipped with this rifle but it is known that they were issued to troops in the Washington, D.C., area.

Fig. 1-3. A .58-60 rimfire cartridge (left) shown with a .50-70 centerfire cartridge (right) for comparison.

Adapted From

The parent arm for conversion to the 1865 Rifle was, oddly enough, not the latest Model 1864 rifle musket, which was being held in reserve while the new breechloaders were being fabricated, but the earlier Model 1861 style, with its distinctive flat barrel bands, and "tulip"-headed ramrod.

Table 1 U.S. Model 1865 .58 Caliber Springfield Rifle	
Finishes	
Receiver	Not Applicable - Strap-Blackened*
Breechblock or Bolt	Blackened* but Color-Case-Hardened breech-blocks have also been observed
Hammer and/or Lock Plate	Color Case-Hardened
Barrel	National Armory Bright
Furniture	National Armory Bright Rear Sight: Browned (blued)
Stock	Oil-finished American Black Walnut
Markings	
Receiver	Not Applicable
Breechblock or Bolt	None
Hammer and/or Lock Plate	Hammer: knurled in shield pattern Lock Plate: “1865,” eagle clutching arrows and olive branch, “U.S./SPRINGFIELD”; dates 1862, 1863, and 1864 have also been observed
Barrel	None (portion of a date and various single-letter subinspector’s stamps from period of use as a percussion musket may be visible near the tang)
Furniture	Butt Plate: “U.S.” on tang Bands (3): “U” near upper edge, right side Rear Sight: “3,” “5” on leaf
Stock	“SWP” in rectangle, and “ESA” in oval on left flat
Principal Dimensions	
Overall Length	56 inches
Stock Length	52-3/4 inches
Barrel Length	40 inches overall, 37-5/8 inches in bore; 3 grooves, 1 turn in 72 inches to the right
Muzzle Diameter	0.775 inch

Ramrod Length	39-7/16 inches (flush with muzzle when stowed)
* The term "blackened" is used to describe the color achieved by case-hardening in oil and water. The oil was floated on water. When the part to be case-hardened was removed from the oven, it was immediately immersed in the oil-water bath. The oil produced the black color and the water hardened the steel.	

Identifying Features

Stock

The original 52-3/4-inch-long musket stock was retained, but much additional inletting was required, below the breech area for the ejector pin arm and its mounting bracket, see Figure 1-4, arrow. Considerable cutting was done both behind and forward of the lock plate. This was performed for such added parts as the extractor slide and its 3-3/4-inch-long tubular spring housing, made of nickel-plated brass, which lies concealed just below the top edge of the stock, parallel to the bore, see Figure 1-5.

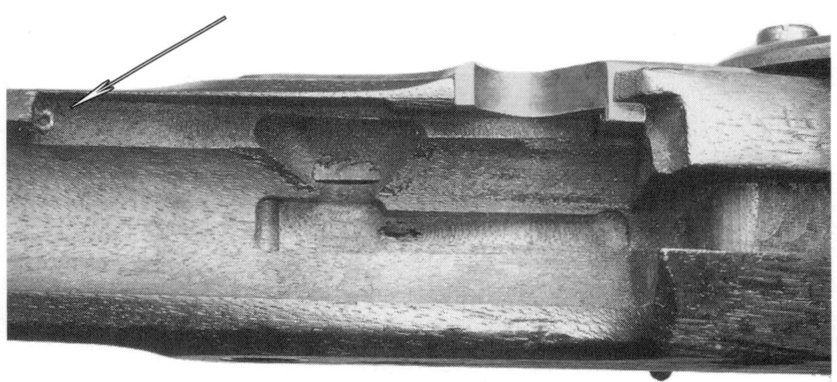

Fig. 1-4. This view of the M1865 inletting shows how wood was removed for the extraction/ejection mechanism. The tiny loop (arrow) receives a pin on the bottom of the extractor slide.

Unfortunately for today's collector, some of these cuts resulted in certain areas of the wood being left only a few hundredths of an inch thick, and many otherwise quite attractive arms are found with

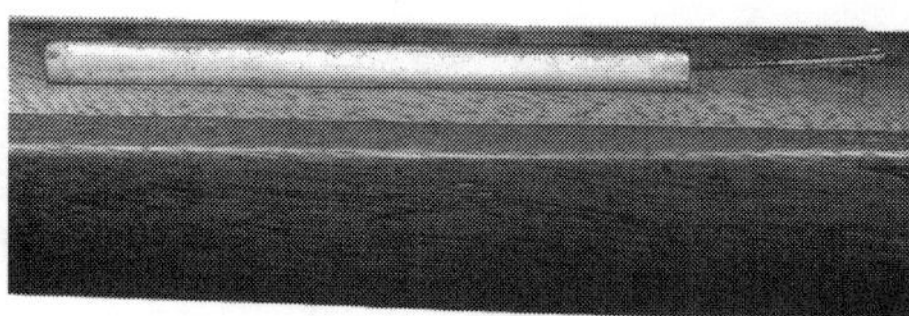

Fig. 1-5. The spring housing in the barrel channel, viewed from left side. A tiny pin on the bottom of the extractor slide engages the loop visible in Figure 1-4 (arrow) on the spring.

unsightly gaps where that wood has broken away, over time. The original three barrel bands were retained, and the shoulders machined into the stock's forend were 11-5/8 inches apart.

Most stocks will have two cartouches: on the left flat, "SWP" (for Samuel W. Porter) in a rectangle with rounded ends, and "ESA" (for Erskine S. Allin) in a small oval, see Figure 1-6.

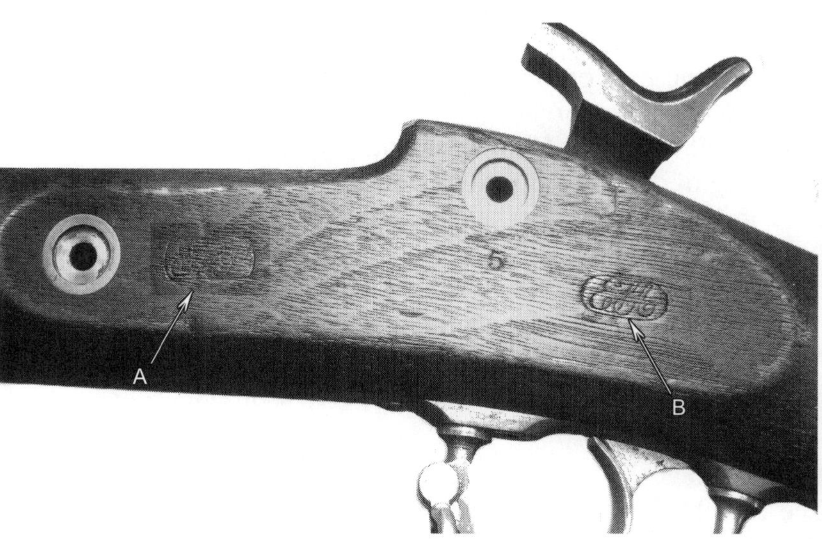

Fig. 1-6. Typical Model 1865 Ordnance Department inspection cartouches on left stock flat, "SWP" (Samuel W. Porter) and "ESA" (Erskine S. Allin), arrows A and B, respectively. Cartouches have been enhanced for clarity.

Barrel

The original U.S. Model 1861 rifle was reused and modified by milling an opening 0.768 inch wide by 1.820 inches long by 0.325 inch deep in the top surface, 0.468 inch ahead of the tang screw joint. Both ends of the cut were angled (when viewed from the side). Length, in the bore, is 37-5/8 inches. Measured along the top of the barrel from joint with the tang screw to the muzzle, the length is 40 inches.

This permitted the insertion of a breechblock, hinged at its front end, which would swing up and expose the rear of the chamber for the new self-contained copper-cased cartridge, see Figure 1-7. This chamber was reamed into the bore, from the rear, removing a portion of the rifling, just ahead of the newly cut top opening, at the same time that space was being provided for the new breechblock. The caliber of the Model 1865 rifle was .58 inch, and the original percussion rifle musket rifling of three broad lands and grooves, with a twist of one turn in 72 inches, remained unchanged.

Fig. 1-7. Left-side view of breechblock. Note the machined recess for the cam-shaft retaining cap (arrow).

Many converted Model 1865 barrels show remains of the last digit(s) of their original date of manufacture, such as "(186)1," "(186)2," "(186)3," or "(186)4." The original percussion rifle musket proof marks may also still be visible behind the new breech opening.

Breechblock

The breechblock was a newly manufactured part and demonstrates clearly the intricate machining need for its complicated components.

There were two distinctly different forms of sighting notch, found on the upper front surface, just to the rear of the hinge "knuckle." On a very few early arms, such as the author's specimen, the first type will be a squared cut approximately 0.410 inch wide by 0.265 inch long, see Figure 1-8, arrow. On later manufactured rifles, the second type was a simple scalloped, or "half-round" cut, see Figure 1-9, arrow, very similar to that which appeared later on the .45-caliber arms.

Fig. 1-8. Top view of breechblock, showing the early squared cutout (arrow) leading to the sighting notch. It is found on only a few specimens. Author's collection.

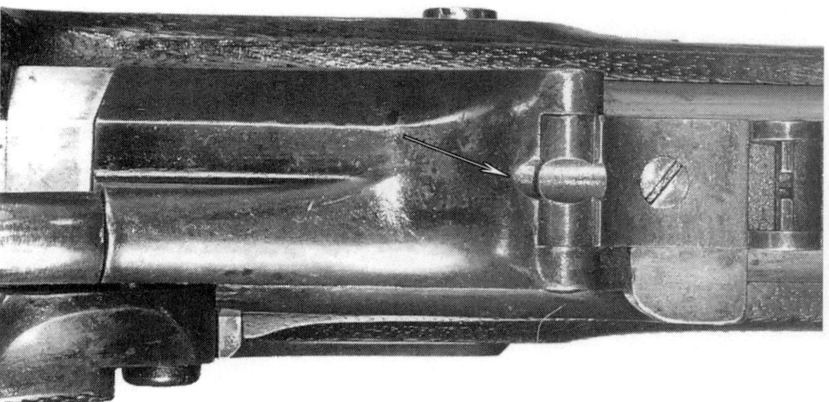

Fig. 1-9. Top view of another breechblock, showing the later rounded cutout (arrow) leading to the sighting notch found on nearly all specimens of the Model 1865. North Cape Publications collection.

Contrary to some previously published works, these blocks *were not* marked in any way! The stamping of a date, the word "MODEL," and the familiar eagle head, crossed arrows, etc., all came later (and not all at the same time).

NOTE: I'm sure that many older collectors will remember that S & S Firearms, the well-known parts house for many years, advertised Model 1865 blocks (probably never a really big mover!) having a mysterious "small chip out of the side." I always wondered why, but now believe this was the notch which engages with, then suddenly releases, the spring hook attached to the sliding extractor rack, thus allowing it to snap forward.

Lock Plate

These were, in most cases, newly made for the rifle, and were dated "1865." The pattern was identical to that of the U.S. Model 1861 and 1863 rifle musket lock plate. U.S. Model 1865 lock plates with dates of 1862, 1863, or 1864 are known in the Springfield Armory Museum collection and have been reported by a number of collectors. It is not known if these other dated lock plates (which account for as many as 26% of those reported in a recent survey) were used on new rifles, placed on experimental arms, or were used as replacements for repaired or rebuilt arms.

Fig. 1-10. Inside view of the lock plate showing the relief milling (arrow) which was added early in production to ensure free operation of the extractor slide.

The distinguishing feature on the true U.S. Model 1865 lock is the relief 0.022 inch deep and 0.520 inch long on the inside of the plate which was needed to provide operating clearance for the extractor slide, see Figure 1-10, arrow. Without this clearance, it is possible to bind the action by overtightening the lock plate screws. On the downside, the resulting gap also allowed water to infiltrate between the metal and wood. The collector should check carefully any U.S. Model 1865 lock plate. On some plates with other dates, this relief cut is either absent or the width and/or depth may vary from the arsenal dimensions. Some even show signs of hand filing. This is not Springfield Armory work.

The lock plate was also marked with the "eagle clutching arrows and olive branch" and "U.S./SPRINGFIELD."

Thumbpiece and Cam Latch

A massive thumb latch, see Figure 1-11, arrow A, attractively checkered on its outer edge, was located at the right rear side of the breechblock, below the hammer. It was fitted into the recess where the drum and nipple of the percussion musket had been, arrow B. No alteration of the lock plate was required to accommodate this part. The area of the stock shown at arrow C, however, was thinned to receive the breechblock and ejector mechanism, so be very careful in this area.

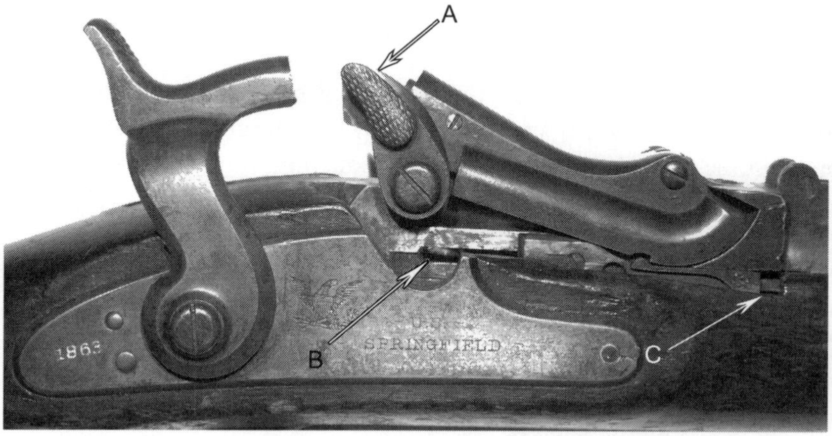

Fig. 1-11. In this side view, note the checkered thumb latch (A), the notch in the lock plate (B) where the percussion nipple and drum had been, and the area of the stock (C) under the front of the breechblock that is very thin. North Cape Publications collection.

The latch assembly, shown in Figure 1-12, consisted of a thumbpiece (A) which was attached to and rotated the cam shaft (B) which ran horizontally through the rear of the breechblock, at a right angle to the bore. A small rearward projecting lug (C), which slipped onto the shaft as a separate piece, was rotated by raising or lowering the thumb latch, in and out of a semicircular recess newly milled into the breechscrew which projected into the rear of the receiver. The removable cap (D) on the left side of the block was held in place by two screws

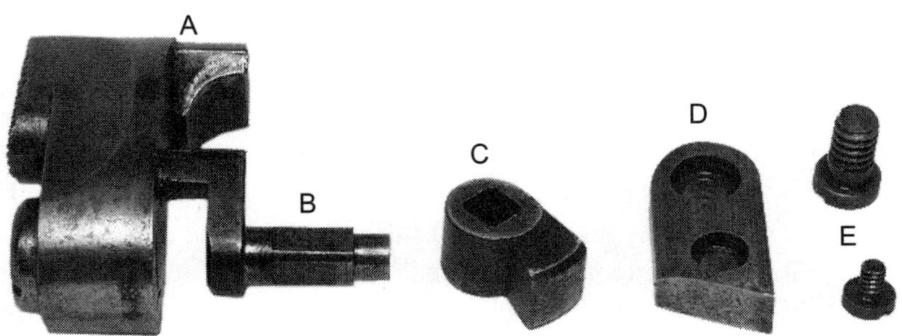

Fig. 1-12. Disassembled view of the breechblock latch: (A) thumbpiece, (B) cam shaft, (C) lug, (D) removable cap, (E) screws. Note also the faceted top of the thumb latch.

(E). The removable cap provided access to the cam-shaft recess. Unlike most later trapdoors, the thumb latch of this model was screwed to its shaft. A very few early guns used a spanner-head screw with two round holes for the spanner wrench, but a large slotted screw was used in the majority of Model 1865 production.

Breechblock Support

The support that held the breechblock on a hinge was made in two pieces, see Figure 1-13. It was 1.215 inches long by 1.061 inches wide, overall. The main, forward section (arrow A) was 0.655 inch long and mounted in the old rear sight dovetail where it was secured to the barrel with one screw and soft solder. The after part (arrow B) was an irregularly

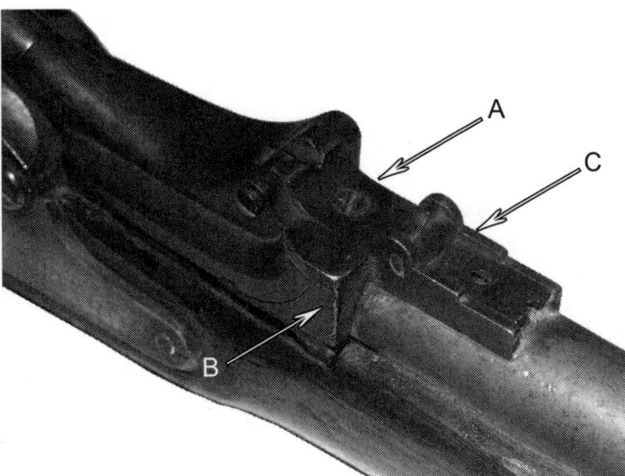

Fig. 1-13. The breechblock support system consisted of two parts: (A) the hinge and (B) the support plate. The rear sight (C) was located flush against the new breechblock and secured only by soft solder.

shaped plate, screwed to the right-hand side of the support where it served as a stop, or closure, for the front of the large upturned flange on the right side only, of the breechblock.

The unique feature of this particular style of breechblock support is that it formed the *center* portion of the hinge joint. In *all* subsequent arms (even the U.S. Model 1866 rifle which also had an applied strap), the *block* formed the central member.

Rear Sight

The original Model 1861 rifle musket rear sight was reused. It was moved slightly forward and located flush against the front of the new breechblock support, refer to Figure 1-13, arrow C. It was fastened to the barrel only with soft solder; a screw was not used, as with later models. The rear sight had one leaf with an aperture for 300 yards plus a notch in the top of the leaf for 500 yards. The sight leaf rotated forward to lie flush.

Hammer

The hammer was altered from the Model 1863 rifle musket percussion hammer by removing the cupping, or cavity, in the head which had covered the percussion cap and nipple when forward, see Figure 1-14. The neck was reshaped to create the pronounced offset necessary to strike the angled firing pin. The offset of the original rifle musket hammer, from bearing face to the center of the head, was approximately 0.740 inch, whereas for the new arm,

Fig. 1-14. The hammer was altered from a U.S. Model 1863 rifle musket percussion hammer by removing the cupping (at arrow) that covered the percussion cap and nipple when down.

the dimension was approximately 0.910 inch, an increase of approximately 0.170 inch. Some very early examples of this arm used a hammer with extended nose, perhaps in an attempt to shorten lock time. However, this proved unsatisfactory due to a lessened striking force and so was reduced to the standard dimension of 1.585 inches, from rear of thumbpiece to nose.

Extracting/Ejecting Mechanism

This was a very complicated system! Teeth, internally milled into a curved gear along the right side of the breechblock, meshed with a sliding rack mounted to the right side of the barrel in a trough behind the chamber, see Figure 1-15.

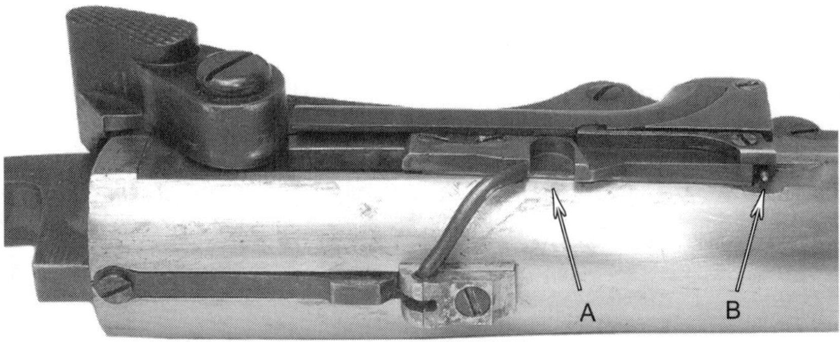

Fig. 1-15. This view of the extracting and ejection mechanism, when viewed from beneath with block closed, clearly shows the slide (A) forward and the ejector mechanism at rest. The pin which engages the loop is shown at (B).

The rack visible when the block was opened had a lip at the front that projected into the chamber, under the head of the cartridge. As the breechblock was lifted open, the rack was drawn back by the internal gearing. At the same time, a coil spring, located in a brass tube next to the barrel, was compressed. The coil spring had a looped end which slipped over a tiny tapered pin at the front of the rack slide.

The roughly "U"-shaped arm (A) seen in Figure 1-16, clamped to the bottom of the barrel, wrapped forward, up and around, to fit into a recess in the slide. This recess in the slide (B) captured the "U"-shaped

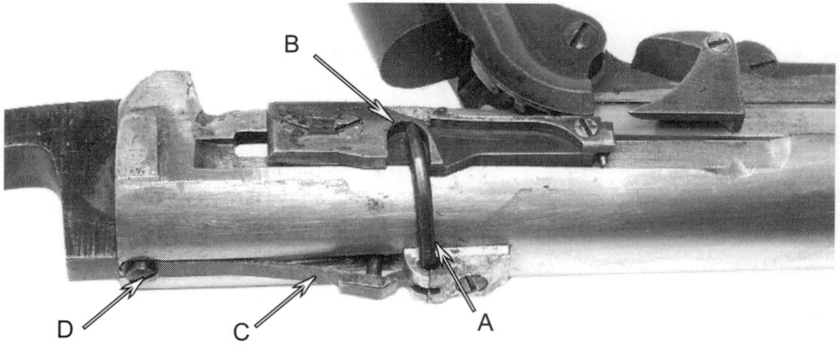

Fig. 1-16. With the breechblock more than half open, its teeth are nearly ready to slip out of the slide which has captured, and partly rotated, the ejector arm. The pin is ready to pop up. Parts are: (A) arm, (B) cam recess in the slide, (C) spring arm, and (D) spring arm attachment point.

arm and caused it to rotate, and, by means of a projection at its lower end, camming down a spring arm which lay in a slot (C) milled into the bottom of the breech.

This spring arm was fastened at its rear end with a small screw (D) immediately in front of the tang. The spring arm had an upward projecting ejector pin at its front end, which protruded through a 0.125-inch-diameter hole into the milled cartridge trough.

Fig. 1-17. Arrows point to (A) the part of the slide that actually touches the cartridge case, and (B) the "pop-up" ejector pin, in the bottom of the receiver trough.

When the breechblock was rotated up approximately 60 degrees, the slide was suddenly disengaged from the teeth on the block, and two things occurred simultaneously as shown in Figure 1-17. The cartridge case, which had been dragged to the rear, was suddenly freed as the extractor slide (A) snapped forward, and the ejector pin (B), whose mounting spring had slipped off its cam, popped up through the bottom of the receiver to throw the empty case out of the breech.

Did you follow all that? Those who would malign the comparatively simple and sturdy post-1867 system should understand how truly complicated its forerunner was!

Furniture

The balance of parts, including the butt plate, trigger guard assembly, swivels, barrel bands, band springs, tip, ramrod, etc., were taken from the parent arm, a standard M1861 Rifle Musket. The butt plate was marked "U.S." The three bands were each 0.610 inch wide by 0.062 thick. They were marked "U" on the right side at the front edge. They appear to be flat in cross section, see Figure 1-18, but actually taper by approximately 0.020 inch from front to rear. The middle band carried a riveted sling swivel.

Fig. 1-18. Close-up view of the flat middle band, unchanged from the parent Model 1861 Rifle Musket.

Ramrod

The slightly tapered rod (0.198 inch at the small end, and 0.228 inch below the tip) had a 0.536-inch "tulip"-shaped head, with a pronounced swell of 0.328 inch centered 5-1/4 inches below it, see Figure 1-19. Below the swell the diameter was also 0.228 inch. The stock was recessed to receive this wider portion; this was the least effective method of retaining the rod on any of the Springfield arms. No spoon spring was used on this model. The rod was 39-7/16 inches long overall; the small end was threaded for the implements (jag, ball puller, etc.) of the parent musket, which, of course, were no longer

Fig. 1-19. Right-side view of muzzle area of the full-length rifle, showing the flat upper band and the "tulip-head" ramrod.

required when it was converted to fire a metallic cartridge. There was no slot for a wiping rag. The ramrod was flush with the muzzle when stowed.

NOTE: The bayonet was considered by the Ordnance Department an integral part of the rifle. One bayonet was packed with each rifle in the shipping crate, and one bayonet was issued with each rifle. But because many of the rifles described in this text are similar, all bayonets issued with each model are described in Chapter 20.

Chapter 2: The U.S. Model 1865 .58 Caliber Springfield Cadet Rifle (aka Model 1865 Rifle, Two-Band Variation)

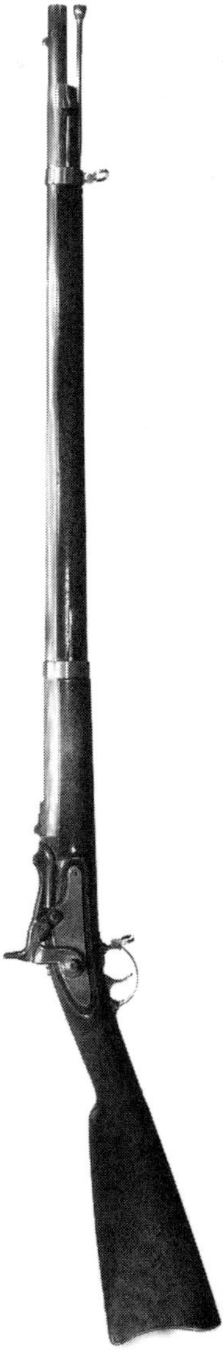

Historical Background

This model continued a practice which began with the miniaturized percussion Cadet Musket of 1851: that of providing an arm of reduced size and weight, for use by the cadets at West Point. However, at this point it should be noted that there is some disagreement among advanced collectors as to the authenticity of this arm as a Springfield product, due to a lack (at least so far) of "official" documentation.

Yet, several of the arms are known to exist, all with the identical special features noted below, including the proper markings, see Figure 2-1. Most importantly, these rifles are in a form distinctly different from the known "surplus-dealer" style of alteration. Surplus-arms dealers such as Bannerman, et al., frequently, if not always, shortened the barrel and stock so that the middle barrel band remained in its original position; they discarded the upper band entirely. This left a very ungainly-appearing rifle. They usually thinned the wrist quite severely, as well. Guns showing these features are simply mutilated Model 1865 rifles and not scarce cadet rifles. Their value is a great deal less.

It should be stressed that the specimen described here is a perfectly shortened version of the 3-band arm.

Fig. 2-1. The U.S. Model 1865 .58 Caliber Springfield Cadet Rifle, also known to some collectors as "First Allin Alteration, two-band variation."

There are no "discrepancies" whatsoever. The cadet rifle is exactly 4 inches shorter overall, and all component dimensions agree, exactly. There are no aberrations as to markings or finish. The machined features, such as the milled lock plate relief, only recently brought to light, are all present and correct.

Quantity Produced

An estimated 270–300 of the Cadet rifles were manufactured, presumably during 1866. A letter from Springfield Armory, dated June 27, 1866, referred to a request for 300 cadet arms and stated that they could be ready in two months. Since this date preceded known production of the Model 1866, it is assumed that Model 1865–type arms were meant. The writer of the letter suggested the substitution of some shortened muzzleloaders, which were on hand for immediate delivery, but no final determination regarding that proposal is known.

Adapted From

As the .58-caliber cadet rifles were a shortened, but properly proportioned, version of the standard (3-banded) Model 1865 Rifle, it is not known whether these arms were fabricated from previously converted full-length 1865 Rifles, or were themselves directly converted, at the shorter length, from the parent muzzle-loading muskets. Most appear to have used the U.S. Model 1865–dated lock plate, see Figure 2-2.

Fig. 2-2. Lock plate of the U.S. Model 1865 Springfield Cadet Rifle dated "1865." Markings enhanced.

Table 2 **U.S. Model 1865** **.58 Caliber Springfield Cadet Rifle**	
Finishes	
Receiver	Not Applicable - Strap-Blackened*
Breechblock or Bolt	Blackened*
Hammer and/or Lock Plate	Color Case-Hardened
Barrel	National Armory Bright
Furniture	National Armory Bright Rear Sight: Browned (blued)
Stock	Oil-finished American Black Walnut
Markings	
Receiver	Not Applicable
Breechblock or Bolt	None
Hammer and/or Lock Plate	Hammer: knurled in shield pattern Lock Plate: Eagle clutching arrows & olive branch "U.S./SPRINGFIELD" "1865" at rear of plate. Other dates, "1862," "1863," "1864," are known.
Barrel	None (portion of a date and various single-letter subinspector's stamps from period of use as a percussion musket may be visible near the tang)
Furniture	Butt Plate: "U.S." on tang Bands (2): "U" near upper edge, right side Rear Sight: "3," "5" on leaf
Stock	"SWP" in rectangle, and "ESA" in oval on left flat

Principal Dimensions	
Overall Length	52 inches
Stock Length	48-3/4 inches; barrel band shoulders 19-1/8 inches apart
Barrel Length	36 inches overall, 33-5/8 inches in bore; 3 grooves, 1 turn in 72 inches to the right
Muzzle Diameter	0.775 inch
Ramrod Length	35-7/16 inches (flush with muzzle when stowed)
* The term "blackened" is used to describe the color achieved by case-hardening in oil and water. The oil was floated on water. When the part to be case-hardened was removed from the oven, it was immediately immersed in the oil-water bath. The oil produced the black color and the water hardened the steel.	

Identifying Features

Stock

The stocks were shortened to 48-3/4 inches, retapered, and neatly recut for the nose cap. They were fully proportioned around the wrist and not thinned down, as in the "Bannerman" style of alteration, see Figure 2-3. The forends were recut for two barrel bands, the shoulders of which were spaced 19-1/8 inches apart. The middle band spring was removed and the slot neatly filled with a strip of walnut, carefully matched for color and grain, see Figure 2-4.

Fig. 2-3. Note the fully proportioned wrist on the U.S. Model 1865 Springfield Cadet two-band rifle.

Master Armorer Erskine S. Allin's inspection cartouche, "ESA" in a flattened oval, was stamped in the wood at rear of the left-side flat. An additional cartouche for Inspector Samuel W. Porter, "SWP" in a rectangle with rounded ends, was stamped just behind the front lock plate screw, refer to Figure 1-6. At one time the latter cartouche was thought to relate to the shortening operation, but is now considered to be part of the normal inspection process and marking for rifles of both lengths.

Fig. 2-4. The arrow points to the nearly invisible, filled-in band-spring slot, at former middle band location.

Breechblock and Mechanism

Please refer to Chapter 1, text and figures, for a detailed description of the construction and operation of the breechloading conversion.

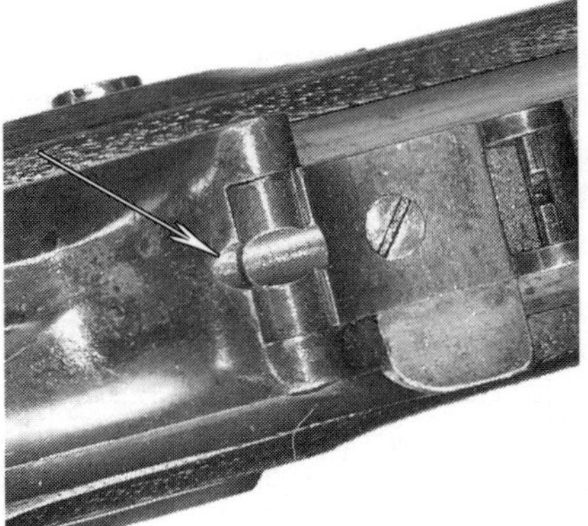

Fig. 2-5. U.S. Model 1865 .58 Caliber Springfield Cadet Rifles have been observed with both the squared and rounded sighting notch. This photo shows the rounded notch on the author's example (arrow).

These two-band guns have been noted with both forms of the sighting-notch variation as described in the preceding chapter. The author's specimen, from which these measurements were taken, has the small rounded notch, see Figure 2-5.

Barrel

The original Model 1861 barrels used on the percussion rifle muskets were shortened to 36 inches (measured from breech-screw joint) or, 33-5/8 inches in the bore. The bore will show a 45-degree internal taper at the muzzle. The barrels were lathe-turned so that they tapered to 0.775 inch at the muzzle, then repolished. This allowed them to accept a standard-diameter M1855 bayonet socket. They retained the standard Model 1861 combination front sight and bayonet stud, remounted 1.213 inches behind the muzzle, the same placement as on the full-sized rifle.

Upper Band

The upper band was moved exactly 4 inches to the rear, see Figure 2-6. The sling ring and base was sawn from the now-discarded middle band and brazed to this relocated upper band. Look closely for the line of yellowish-colored braze.

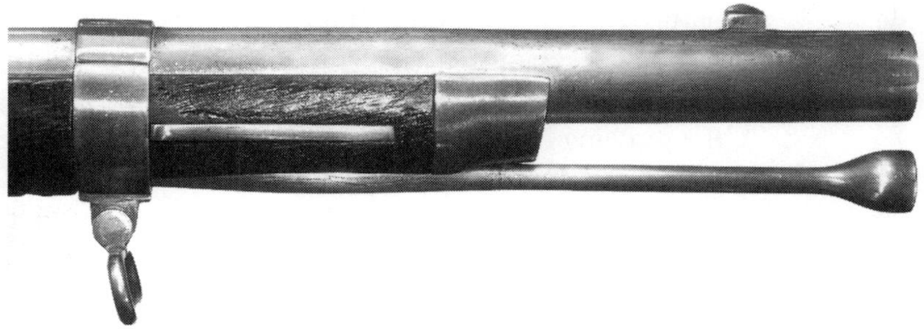

Fig. 2-6. The sling swivel was cut from the (discarded) middle band, and neatly brazed to the upper band. Note that front sight is properly mounted, and that proportions of wood and metal are correct.

Ramrod

The original Model 1861 ramrod was shortened (from the rear) to 35-7/16 inches. The end was neatly rounded over. Since no implements were required to be attached, it was not rethreaded. The ramrod retained the original cupped tulip head and swell to hold it in the ramrod channel. These dimensions were the same as the ramrod used on the full-sized .58-caliber Model 1865 as described in the previous chapter. The ramrod was flush with the muzzle when stowed, refer to Figure 2-6.

Chapter 3: The U.S. Model 1866 .50-70 Springfield Rifle (aka "Second Allin Alteration")

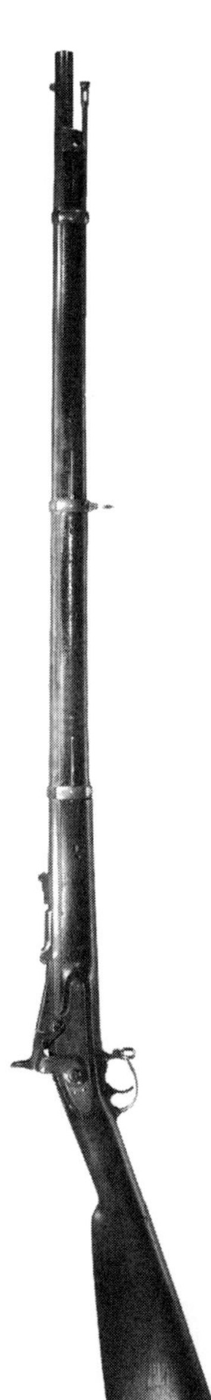

Historical Background

The 1865 conversions had proved to be a disappointment. The cartridge, an underpowered rimfire design with the weak case head inherent to that type, was not considered capable of much real improvement. Also, the design of the action was extremely complicated and had numerous parts, some of them quite fragile. The extracting and ejecting mechanisms were of particularly immature design. At the same time, the technology involving the drawing of metal cartridge cases was improving, and so the decision was made to change to a longer, more powerful cartridge and a simpler and sturdier action design.

The new cartridge was made of copper; it was centerfire, rather than rimfire, and used a new .50-caliber bullet. The powder charge was 70 grains of FFg black powder, and the rather pointed lead bullet weighed 450 grains.

Model 1866 rifles, see Figure 3-1, gave a very good account of themselves at the Wagon Box Fight, and the Hayfield Fight, both of which took place along the Bozeman Trail in Wyoming and Montana Territories in August of 1867.

The former skirmish was particularly noted for the fact that thirty-two soldiers and teamsters soundly

Fig. 3-1. The U.S. Model 1866 .50-70 Springfield Rifle, also known as the "Second Allin Alteration." This is the model used by the Army, with telling effect, at the Wagon Box and Hayfield Fights, both in 1867.

defeated a much larger force under the personal leadership of the famous Sioux chief, Red Cloud, with but relatively minor losses. For years, lurid claims were circulated regarding how many warriors were involved, with some "guestimates" ranging up to 3,000. In truth, the total was likely to have been below three hundred. But even at odds of nearly ten to one, it was still an impressive showing. This victory was due in large part to the new U.S. Model 1866 Springfield Rifle which inflicted significant casualties on the attackers, who expected the soldiers would have to pause to reload the muzzleloading rifle muskets with which they had been equipped in the past.

This very same tactic, attack while the enemy is reloading (plus an outstanding decoy-into-ambush, at which the Sioux truly excelled), had been used successfully in the equally overwhelming Indian victory against the Fetterman command just a few months earlier.

Quantity Produced

Approximately 52,300 of the U.S. Model 1866 Springfield Rifles were manufactured, nearly all during 1867, plus a few hundred more made in both 1868 and 1869 for U.S. service use. Just over 3,100 additional arms were assembled from parts in 1871 during the Franco-Prussian War for sale to France. The Model 1866 was not serially numbered.

After being replaced in U.S. service by the Model 1868, a great many of these arms were purchased by Remington for sale abroad. Many were furnished to the French during the Franco-Prussian War of 1870–1871. Subsequently, many of those found their way back to this country. So, while they are by no means rare, we do not see as many of them today as might be expected based on the numbers manufactured. Also, usage abroad, as well as purchases by surplus-arms dealers, led to many of the rifles having "scrambled" parts. This, of course, makes it difficult for today's collector to say with absolute certainty what was right, and what was not. Research continues to this day. The story of the Model 1866 Springfield, the "Second Allin," is proving to be a very complicated one.

The .58- and .50-Caliber Rifles

Adapted From

Unlike their predecessors, these 1866 "Second Allin" conversions were based upon the last of the rifle muskets made: the U.S. Models of 1863 and 1864, with their narrower, oval-shaped bands, band springs, and ramrod having a seven-ringed, cupped head, made without retaining swell.

Presumably this was due to the fact that many more of that model were available, and with the commitment now firmly made to switch to a breechloader, it was no longer considered necessary that they be held in reserve.

Table 3 The U.S. Model 1866 .50-70 Springfield Rifle	
Finishes	
Receiver	Not Applicable - Strap-Blackened*
Breechblock or Bolt	Blackened*
Hammer and/or Lock Plate	Color Case-Hardened
Barrel	National Armory Bright
Furniture	National Armory Bright (unless otherwise noted) Rear Sight: Browned (blued)
Stock	Oil-finished American Black Walnut, 50% of the rifles issued; the remainder were varnished
Markings	
Receiver	Not Applicable
Breechblock or Bolt	"1866"/eagle head
Hammer and/or Lock Plate	Hammer: knurled in shield pattern Lock Plate: Eagle clutching arrows and olive branch, "U.S./SPRINGFIELD" "1864" at rear of plate. Other dates are known
Barrel	None (portion of a date and various single-letter subinspector's stamps from period of use as a percussion musket may be visible near the tang)

Furniture	Butt Plate: "U.S." on tang Bands (3): "U" near upper edge, right side Rear Sight: "3," "5" on all, and "7" on some leaves
Stock	From one to three cartouches, with various initials, on left stock flat (refer to text). Some rifles will show a script *N* approximately 2 inches behind the trigger plate. A very few rifles will show two *N*s
	Principal Dimensions
Overall Length	56 inches
Stock Length	52-3/4 inches; barrel band shoulders 11-5/8 inches apart
Barrel Length	40 inches overall, 36-5/8 inches in bore; 3 grooves, 1 turn in 42 inches to the right. Barrels were lined
Muzzle Diameter	0.775 inch
Ramrod Length	38-3/4 inches (1-1/8 inches behind the muzzle when stowed)
* The term "blackened" is used to describe the color achieved by case-hardening in oil and water. The oil was floated on water. When the part to be case-hardened was removed from the oven, it was immediately immersed in the oil-water bath. The oil produced the black color and the water hardened the steel.	

Identifying Features

Stock

The stock may have been either reused from the parent musket, or newly made. In either case, additional cuts, see Figure 3-2, arrow, were required for the ejector spring and the friction spring arm, see Figure 3-3, which activated the pop-up pin-type "cartridge retainer" that was located just behind the chamber (see description of extracting process below).

The cartouche(s) on the left stock flat, see Figure 3-4, will indicate

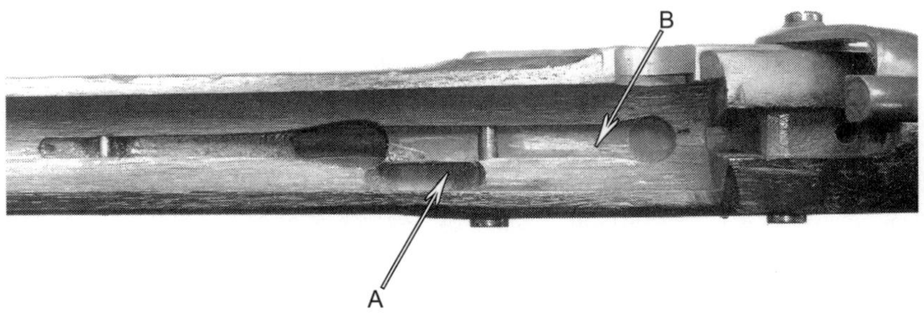

Fig. 3-2. Additional inletting was required for the "horseshoe" ejector spring (A) and the friction spring (B). Note spoon pin inletting, and pin, but the spoon itself was removed, prior to use.

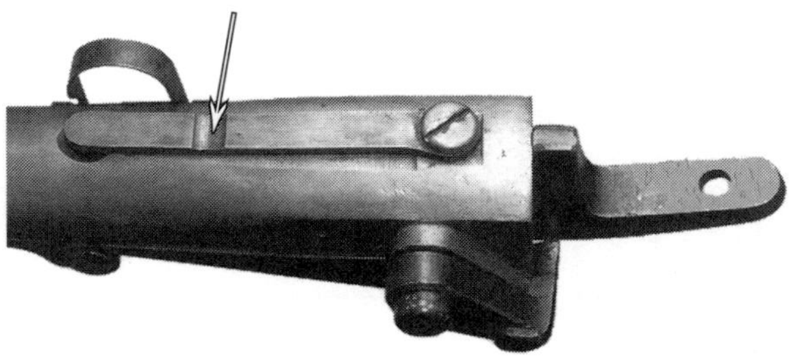

Fig. 3-3. This long spring, which was notched for the front lock screw (arrow), powered the pin which kept the cartridge from backing out of the chamber, once inserted therein. It was referred to as the "friction spring" by the Ordnance Department.

whether the stock was new or re-used. An old stock may have as many as three stamps, whereas a new one will have but one, frequently "HSH," in a small (0.670 inch by 0.245 inch) box with oval ends. Often, there was no "ESA" cartouche even though Erskine S. Allin was the master armorer at the time. Research is still ongoing to better determine the scope and meaning of the multiple cartouches found on arms of the 1865–1870 period.

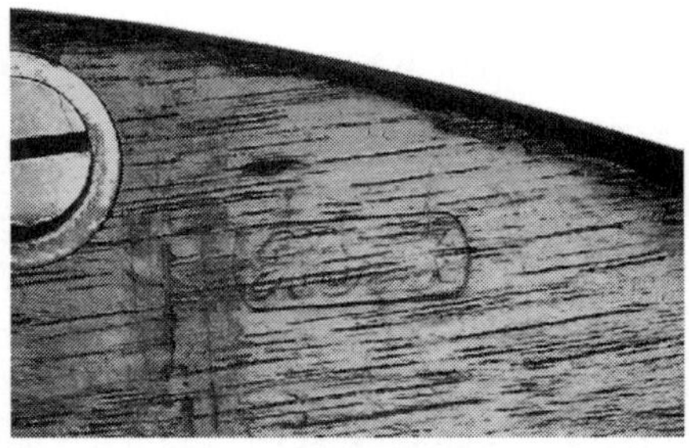

Fig. 3-4. Ordnance Department final inspection cartouche on left stock flat. Several different configurations are known.

Stock length was 52-3/4 inches, including butt plate and tip. There were three bands, equally spaced; the shoulders were 11-5/8 inches apart.

Barrel

The existing .58-caliber musket barrel from the Model 1863 rifle musket was used for the Model 1866 rifle. It was 40 inches long overall (36-5/8 inches in the bore) with a muzzle diameter of 0.775 inch. It was reamed out to 0.640 inch for its full length, fitted with a .50-caliber liner, and crowned at the muzzle, as were all future barrels. The liner was rifled with three broad lands and grooves of equal width. The liner was carefully brazed in place at muzzle and breech. The brazing operation was done so well that the author has been unable to find any reports of the lining coming loose in service. The drum was drilled out for the thumb-latch shaft, and a small filler was welded in to flush-off the recess where the nipple had been screwed.

The rate of rifling was increased in this model from that of the parent musket, to one turn in 42 inches. The cutout at the top of the barrel, for the breechblock, was longer and narrower than that of the U.S. Model 1865 .58 Caliber Springfield Rifle at 0.710 inch wide by 2.930 inches long. Metal was also removed on the left side, at the front edge of the block cutout, to form a small flat, see Figure 3-5, which would accept the new ejector spring, which was retained by a cap plate and screw.

Breechblock

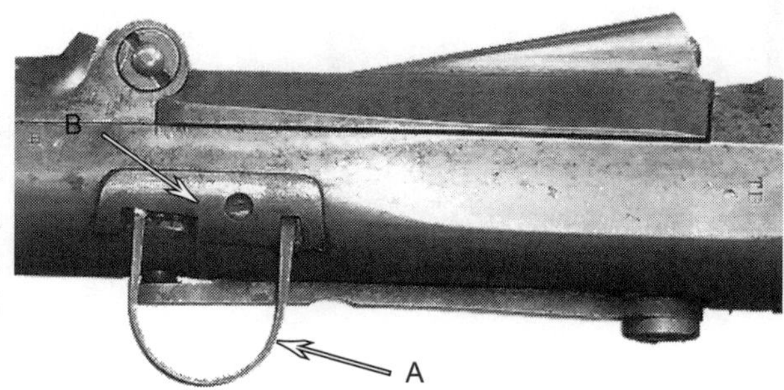

Fig. 3-5. Left side of barreled action, showing the "horseshoe" ejector spring (A) and the retaining plate (B).

The breechblock was longer and narrower than that used on the U.S. Model 1865 .58 Caliber Rifle. Early blocks used on the Model 1866 had a round-topped firing-pin housing (comb), while later ones had a distinct flat on the top, see Figure 3-6. The right-side edge (of both styles of comb) ran straight for nearly the entire length of the block, unlike later arms in the trapdoor series in which it stopped well short of the hinge. This difference was most noticeable when viewed from the right (lock) side, see Figure 3-7. All breechblocks were marked on the top with the date "1866," and a very small eagle head.

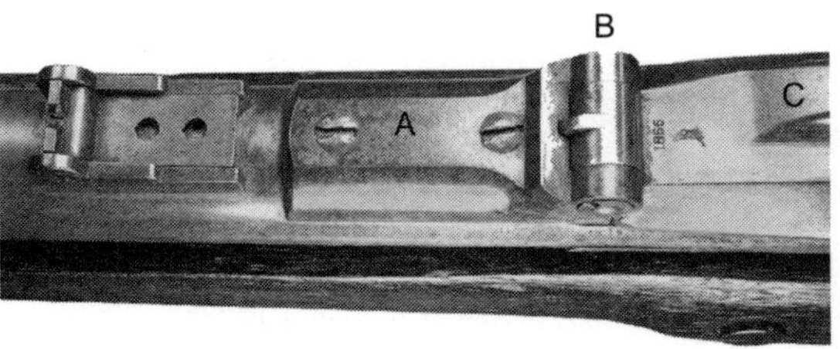

Fig. 3-6. Breechblock attachment (A). Note the lack of a sighting notch on the hinge pin housing (B), and the later, flat firing-pin housing (C).

Alone in the entire Allin series, the Model 1866 breechblock did not have a sighting notch at the hinge. Also unique was the fact that it swung open 160 degrees, far more than any other version of the trap-door action. There was no "arch" milled into the bottom of the breech-block, as was found in later models; the straight cylindrical profile was interrupted only by a 0.280-inch clearance hole for the ejector stud, see Figure 3-8.

Fig. 3-7. Right-side view of Model 1866 breechblock showing the hinge strap attachment, and the rear sight. Note the lock plate markings.

The Model 1866 breechblock was configured for the centerfire .50-70 cartridge and used a long, spring-loaded firing pin to strike the primer. The firing pin was retained by a threaded collar at its rear end, removable with the spanner-head screwdriver on the Model 1866 combination tool described in Chapter 19. This new and much longer firing pin gave rise to the widely used nick-name of "needle gun," which was applied to any Allin "trapdoor" commencing with this 1866 model. Do not confuse this "American" term with the much older Prussian needle-fire *system*, with its special front-ignition paper cartridge!

The thumb latch was reshaped; it was slightly smaller than that of the 1865, but was still noticeably wider than the one found on later arms, refer to (B) in Figure 3-8. It was no longer checkered along the exposed top edge, but it was still removable from the shaft, being fastened with a large, flat-head, slotted screw.

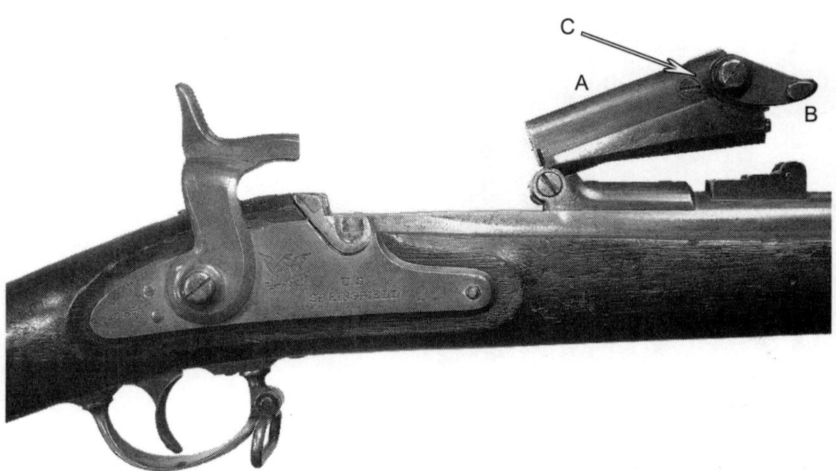

Fig. 3-8. The breechblock opened a full 160 degrees, more than any other Allin-type arm. Note absence of an "arch" on the bottom (A), the smaller thumb latch (B), and retaining plate (C) indicating the second style of block.

The thumb latch was made with two distinct forms of cam-latch attachment; the earlier pattern was much like that of the 1865, with the thumbpiece shaft entering from the right side, and square shaped at its midpoint to accept the small removable (slide-on) cam. The shaft was retained in the block by a small screw entering its end from the left side. Approximately 7,000 rifles (under 15 percent of the total production) were made with this style of block and cam shaft, so they are a desirable (but often overlooked) variant.

The later (and much more often encountered) type used the familiar combination thumb latch and shaft with integral machined cam, retained by a cap plate and screw, let into the right side of the block, refer to Figure 3-8, arrow C. This was the same general design as that used on all subsequent models, through the end of production of the Allin-patent rifles in 1893; the only difference was that here, the thumbpiece was screwed, rather than staked, to the shaft.

Like the U.S. Model 1865 .58 Caliber Rifle before it, the breechblock pivoted on a screw, which was originally retained by a single slotted nut on the left side, see Figure 3-9, arrow A. Problems in service led to eventual use of a two-part jam nut, which protruded slightly from the breechblock on the left side, refer to the inset in Figure 3-9.

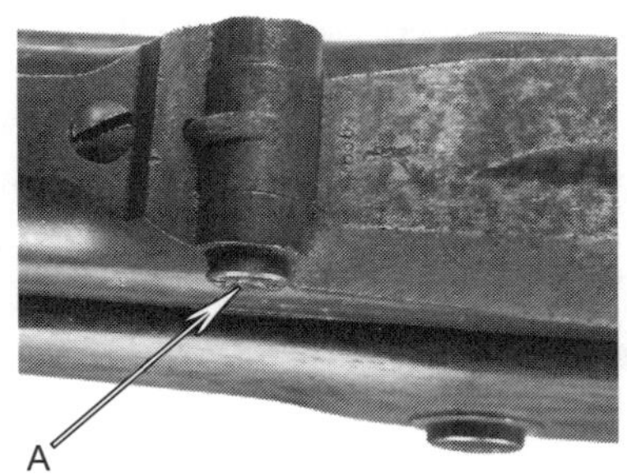

Fig. 3-9. Note breechblock hinge screw nut (A) and extractor (B) on the breech-block. A later version (inset) used a longer hinge screw and double locking nuts.

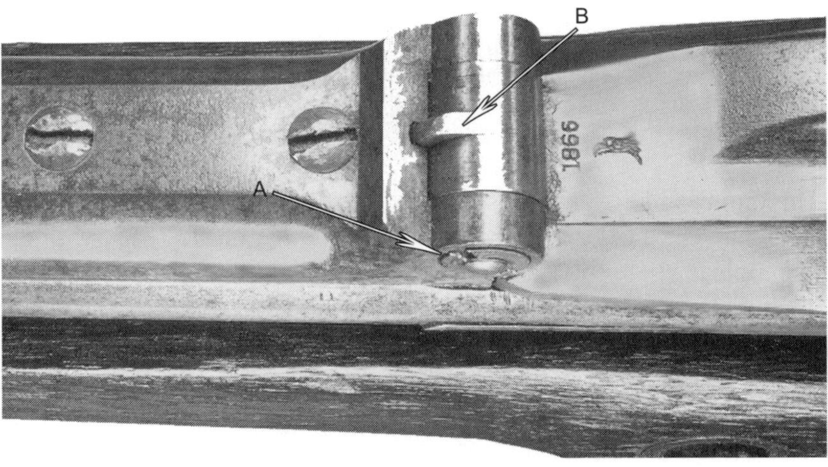

Extractor

Primary extraction was by means of the small "hook" projection on the top front of the breechblock hinge knuckle, refer to Figure 3-9, arrow B, which caught the rim of a chambered case, as the block was rotated open.

Final extraction and ejection was powered by a horseshoe-shaped spring, fastened by a plate and screw to the lower left side of the barrel, see Figure 3-10. A small lug on the front arm of this spring projected into the receiver just behind the rim of the chamber; thus the spring was extended, or spread apart, by the cartridge case as it was being chambered.

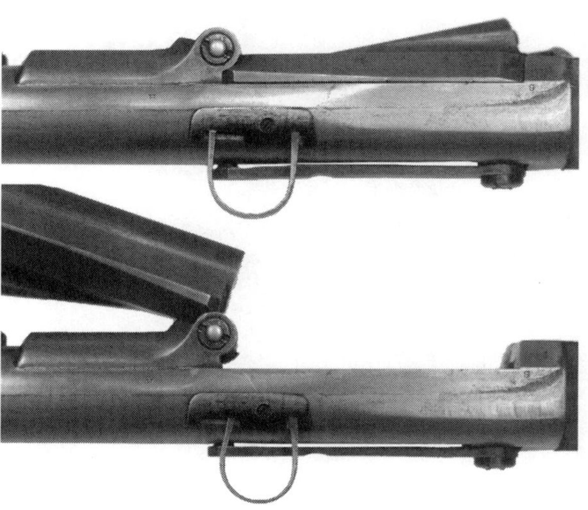

Fig. 3-10. This composite photograph shows the action of the extractor spring when the breechblock is closed (top) and open (bottom).

A small spring-loaded pin, beveled on its rear face, rose from the front of the receiver trough, just behind the chamber, to prevent escape of the cartridge after it had been inserted against the pressure of the horseshoe ejector spring.

When the breechblock was rotated open, the "hook" forced the fired case past the spring-loaded pin. The extractor/ejector spring, which was now contracting, pulled the case back to strike the angled face of the ejector stud mounted in the floor of the receiver and deflected it up and out. This was a much better system than that used in the Model 1865 .58 Caliber Rifle, but it depended on a fragile leaf-type machined spring, and was still considered unnecessarily complicated.

Hammer

The hammer nose was slightly angled, as is the rear of the breechblock comb, so that the joint between them did not form a line perpendicular to the bore, as would be expected, see Figure 3-11, arrow. The reason for this unusual feature, observable only from the top, is not known.

Fig. 3-11. Junction of the comb, or hammer and firing pin housing (arrow). Note that they do not come together at a right angle to the bore, as do all other all other Allin arms.

Rear Sight

The rear sight base was not soldered in place against the hinge strap as it was in the Model 1865 .58 Caliber Rifle, but was moved forward approximately 0.270 inch. It was secured in a dovetail slot with a spanner-head screw, as on the original Model 1863 and 1864 Rifle Muskets. But in the Model 1866, the sight was reversed so that the leaf, which had two fixed sighting holes (for 300 and 500 yards, with a 700-yard notch at top) was now hinged at the front, and folded down toward the receiver, the opposite of the rear sight used in the Model 1865 .58 Caliber Rifle, see Figure 3-12. The leaf screw entered from the left side. All sight leaves will be found with the graduations "3," "5," and some with "7," located on the "wrong" or muzzle side of the leaf. The small mill cuts which form the knife-edges for the sighting notches face the firer when the leaf is erected. The reason for this anomaly, which has been observed on some very fine specimens, is not known.

Fig. 3-12. The U.S. Model 1863/64 Rifle rear sight was installed in reverse on the U.S. Model 1866 Rifle.

Ramrod

The ramrod was 38-3/4 inches long. When properly stowed in the ramrod channel, it was approximately 1-1/8 inches short of the muzzle, see Figure 3-13.

The 0.450-inch maximum-diameter slotted head (having seven 0.396-inch rings) was cupped; shaft diameter was 0.255 inch behind the head, tapering to 0.230 inch at the tip, which was threaded (#12-26) to fit a small steel plate which was inletted into the bottom of the stock, in the trigger plate mortise.

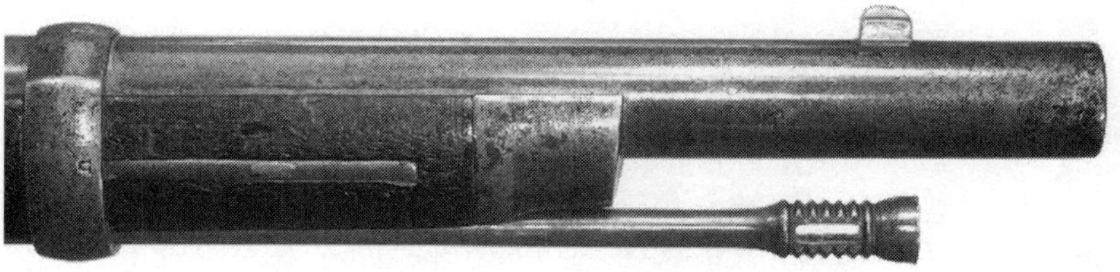

Fig. 3-13. Right-side view of muzzle area, showing upper band, and the set-back of stowed ramrod.

The ramrods used in the U.S. Model 1866 Rifle and the U.S. Model 1866 Cadet Rifle (Chapter 5) were retained by screwing the tip into a threaded plate inletted into the stock, rather than by use of the swell as on the U.S. Model 1865 Rifle, or by friction against the ramrod spoon as in the percussion-rifle models. Almost all of the stocks used for the Model 1866 Rifles retained the spoon mortise; many also kept the pin, even though the spoon itself was removed as it was very close to the mortise for the extractor spring.

Chapter 4: The U.S. Model 1866 .50-70 Caliber "Short Rifle" (aka Model 1866 Rifle, Two-Band Variation)

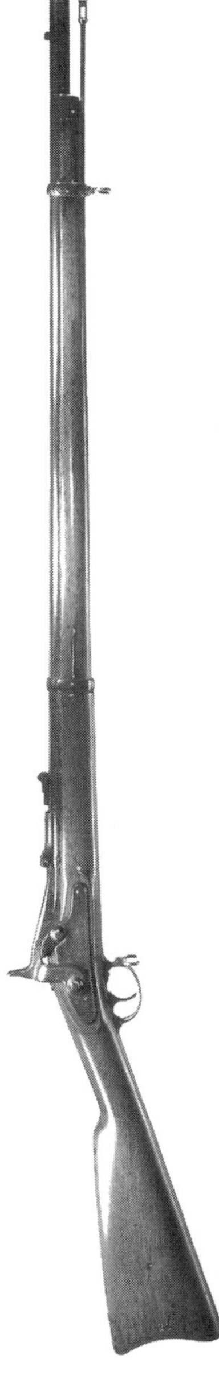

Fig. 4-1. The U.S. Model 1866 .50-70 Caliber "short rifle." (aka Second Allin Alteration, two-band variation). The ramrod shown in this specimen is incorrect.

Historical Background

Something of a mystery arm and again somewhat lacking in "official" status, the U.S. Model 1866 .50-70 caliber "short rifle" has features similar to the U.S. Model 1865 Springfield Cadet Rifle (Chapter 2): it is a properly proportioned arm, see Figure 4-1, but will show signs of having been shortened to its present length.

Correspondence exists (from 1871 to 1872, long after they were general issue) stating that "Model 1866 arms which had burst at the muzzle could be made serviceable by shortening them by four inches"; this process thus made them the same length as the U.S. Model 1868 and U.S. Model 1870 Springfield rifles. These altered guns must not be confused with the Model 1866 Springfield Cadet Rifle (Chapter 5), which was a truly down-scaled or "miniaturized" arm.

All component parts of this two-banded Model 1866 rifle are full size; it is only in overall length of barrel, stock, and ramrod, and number of barrel bands, that they vary from the standard U.S. Model 1866 .50-70 caliber rifle.

Quantity Produced

We do not know for certain the actual number produced, as definitive Springfield Armory production data has not surfaced. One number,

1,146, is mentioned in Armory correspondence, and, in the one brief run of weekly inventory and production reports which have survived (from 1886), 1,019 such arms are noted as being "on hand, ready for issue." Thus they do have somewhat more recognition than the short Model 1865 rifle described in Chapter 2. In this inventory, the rifles are listed as "Springfield Rifles, Model 1866 (Short)."

NOTE: Many Model 1866 Springfield rifles were similarly altered, by the surplus dealers (Bannerman; Schuyler, Hartley & Graham, etc.) of the day; additionally, there is still lack of agreement as to exactly what defines a Springfield Armory–manufactured Model 1866 short rifle. This was very definitely a "second-class" arm, based on damaged weapons that were salvaged.

Adapted From

The short Model 1866 rifle is generally understood to be a further modification of the standard U.S. Model 1866 Springfield rifle with three barrel bands, which had themselves previously been converted to that configuration from the U.S. Model 1863 Springfield rifle musket. It should be noted that the Model 1863 rifle musket model did not have band springs but instead used "clamping" barrel bands; hence when reusing the stock, no band spring filler was needed, see Figure 4-2.

Table 4 The U.S. Model 1866 .50-70 Caliber "Short Rifle"	
Finishes	
Receiver	Not Applicable - Strap-Blackened*
Breechblock or Bolt	Blackened*
Hammer and/or Lock Plate	Color Case-Hardened
Barrel	National Armory Bright
Furniture	National Armory Bright (unless otherwise noted) Rear Sight: Browned (blued)
Stock	Oil-finished American Black Walnut

The .58- and .50-Caliber Rifles

Markings	
Receiver	Not Applicable
Breechblock or Bolt	"1866"/eagle head (small)
Hammer and/or Lock Plate	Hammer: knurled in shield pattern Lock Plate: Eagle clutching arrows and olive branch, "U.S./SPRINGFIELD" "1863" at rear of plate
Barrel	None (portion of a date and various single-letter subinspector's stamps from period of use as a percussion musket may be visible near the tang)
Furniture	Butt Plate: "U.S." on tang Bands (2): "U" near upper edge, right side Rear Sight: "3," "5" on leaf
Stock	From one to three cartouches, with various initials, on left stock flat. Some rifles will show a script *N* approximately 2 inches behind the trigger plate. A very few rifles will show two *N*s
Principal Dimensions	
Overall Length	52 inches
Stock Length	48-3/4 inches; barrel band shoulders 19-1/8 inches apart
Barrel Length	36 inches overall, 32-5/8 inches in bore; 3 grooves, 1 turn in 42 inches to the right. Barrels were lined
Muzzle Diameter	0.775 inch
Ramrod Length	34-3/4 inches (1-1/18 inches behind muzzle when stowed)

* The term "blackened" is used to describe the color achieved by case-hardening in oil and water. The oil was floated on water. When the part to be case-hardened was removed from the oven, it was immediately immersed in the oil-water bath. The oil produced the black color and the water hardened the steel.

Identifying Features

Stock

The stock length, including butt plate and nose cap, is 48-3/4 inches. Some small variation in this dimension will be found. The rifles are

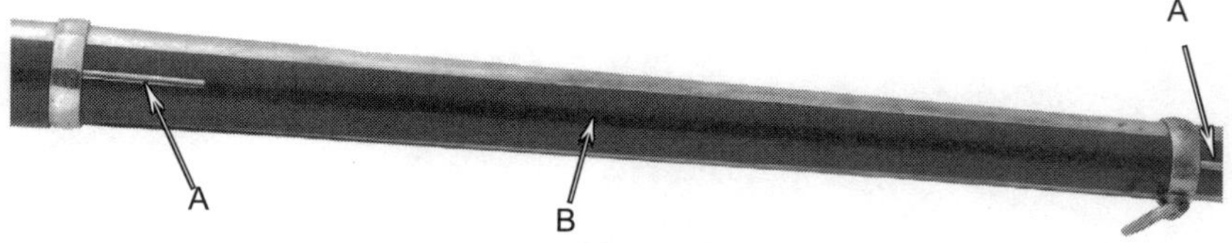

Fig. 4-2. The U.S. Model 1866 .50-70 "short rifle" did not use a middle band. Some specimens will show the middle band-spring slot filled in while others will show rear and forward band-spring slots and band springs (arrows A) but not a middle band-spring slot (arrow B), suggesting they were newly manufactured. Ed Hull collection.

found in the following configurations: (1) utilizing the stock from the U.S. Model 1863 Springfield rifle musket, which did not have a middle band spring slot. This is now thought to be more "correct" than (2) altered from a longer stock, with the middle band-spring slot very carefully filled with a strip of walnut, matched for color and grain, 3) newly manufactured stocks without a middle band-spring slot, refer to Figure 4-2, arrow B. Two bands (arrows A) were used and the shoulders were 19-1/8 inches apart. This is one way to determine if a specimen is "correct" as spacings as close as 18-7/8 inches have been observed on incorrect stocks.

Stocks may, or may not, have the spoon cutout, but the spoon itself was removed as it was very close to the cut for the horseshoe-shaped ejector spring. Some stocks are found with machine-cut inletting for both the horseshoe spring (unique to the U.S. Model 1866 Springfield rifle) and the squared reliefs for the front of the U.S. Model 1868 Springfield receiver; also for the clearance cut for its hinge pin arm.

Barrel

The barrel length is 36 inches, measured from breech screw joint; or, 32-5/8 inches measured in the bore. Very small variations in these dimensions are encountered from rifle to rifle. The barrels were retapered after shortening for the U.S. Model 1855 bayonet. The rear sight was mounted on the barrel in the same manner as the standard-issue U.S. Model 1866 .50-70 Springfield Rifle, see Figure 4-3, arrow A.

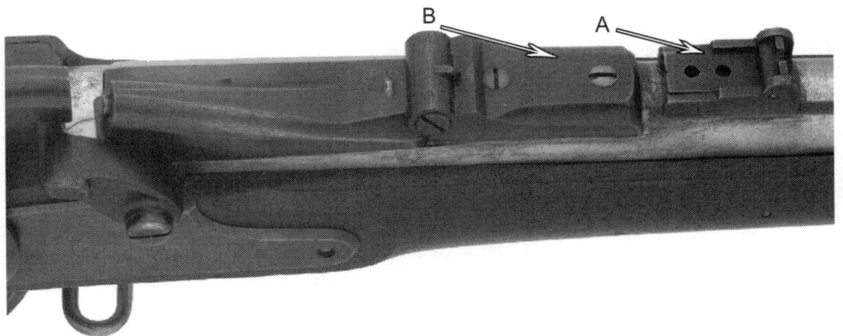

Fig. 4-3. Receiver and breechblock, lock work, and hammer were all reused from the service-issue U.S. Model 1866 rifle. The rear sight was mounted in the same position as on the service rifle (A). Note wider top flat on strap (B). Ed Hull collection.

Breechblock

The breechblock was identical to the breechblock used on the standard-issue U.S. Model 1866 .50-70 Springfield Rifle. The breechblock straps occur in two forms: 1) identical to the original full-length rifle, and 2) of a thicker, less beveled and more blockish shape, suggesting new production of this part during the 1871 period, refer to Figure 4-3, arrow B.

Furniture

This modified arm uses the same butt plate, trigger assembly, lower band, and tip as does the parent U.S. Model 1863 Springfield rifle musket. The upper band, with its sling swivel attached by a screw, is the same as is found on the U.S. Model 1868 Springfield Rifle.

Ramrod

The ramrod used in the short U.S. Model 1866 rifle is presently believed to be 34-3/4 inches in length. This is 4 inches shorter than that of the full-length Model 1866 rifle. The head and its diameter of 0.230 inch remain the same as the long rod. Note that the shortened rods were rethreaded (#12-26) to fit the screw-in ramrod stop, which leaves the rod tip approximately 1-1/8 inches behind the muzzle when stowed, see Figure 4-4.

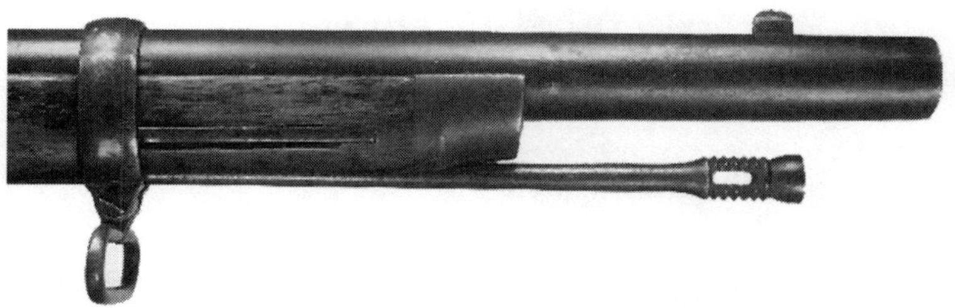

Fig. 4-4. Right-side view of the muzzle area, showing upper band with swivel, and the proper setback of stowed (threaded-in) ramrod, which is 4 inches shorter than that of the full-length rifle.

Note: This is a *very* misunderstood rifle, and the last word concerning exactly what is correct with respect to configuration of the stock, ramrod stop, and the ramrod itself has, in the author's opinion, yet to be written. One mystery is: since the rifles were not made until 1871–1872, why did they (apparently) not use the then-current Model 1868–1870 ramrod style?

Input from other collectors is earnestly solicited.

Chapter 5: The U.S. Model 1866 .50-55 Springfield Cadet Rifle (aka U.S. Model 1867 .50-55 Springfield Cadet Rifle)

Historical Background

An early example of the ongoing series of shorter, lighter arms made for the cadets at West Point, it was officially known as the U.S. Model 1866 Springfield Cadet Rifle and was chambered for the .50-55 caliber cartridge with a 430-grain bullet. However, many people (including the author) prefer to refer to it as the 1867 Cadet. This avoids any confusion with the previously described two-banded variation of the 1866, which was somewhat similar in general appearance, but was not at all the same gun. The special lock plate of this model was dated "1867," the *only* Springfield Armory-produced arm to bear this date.

This is a well-documented, scaled-down rifle, using numerous unique parts not found in any other arm of the Allin series, see Figure 5-1. As such, there are no questions about authenticity attached to it, as there is in the case of the short versions of the Models 1865 and 1866. The Model 1866 (1867) Cadet had been in use only for a short time when they were replaced at West Point by the Model 1869 Cadet Rifle (Chapter 7). Many of these arms were then turned over to the state of Kentucky.

Fig. 5-1. U.S. Model 1866 .50-55 Cadet Rifle, also known as the Model 1867. This rifle was a smaller-scale version of the U.S. Model 1866 .50-70 rifle, designed to be used by military, private, and public school boys. Photograph courtesy of Springfield Armory NHS, SPAR #1129.

Quantity Produced

Just 424 were made: 320 during 1867, and 104 more during 1868. They are very rarely encountered today, and, when found at all, are often in very heavily used condition.

Adapted From

The parent design, and general appearance, except length, was clearly that of the Model 1866 service rifle, but, many parts (detailed below) are not interchangeable.

Table 5 The U.S. Model 1866 .50-55 Springfield Cadet Rifle, aka U.S. Model 1867 .50-55 Springfield Cadet Rifle	
Finishes	
Receiver	Not Applicable - Strap-Blackened*
Breechblock or Bolt	Blackened*
Hammer and/or Lock Plate	Color Case-Hardened
Barrel	National Armory Bright
Furniture	National Armory Bright (unless otherwise noted) Rear Sight: Browned (blued)
Stock	Oil-finished American Black Walnut
Markings	
Receiver	Not Applicable
Breechblock or Bolt	"1866/eagle head" (small)
Hammer and/or Lock Plate	Hammer: knurled in shield pattern Lock Plate: Eagle clutching arrows and olive branch, "U.S./SPRINGFIELD" "1867" at rear of plate. No other date used.
Barrel	None (portion of a date and various single-letter subinspector's stamps from period of use as a percussion musket may be visible near the tang)

Furniture	Butt Plate: "U.S." on tang Bands (2): "U" near upper edge, right side Rear Sight: "3," "5" on leaf (may be on the front or back of the leaf)
Stock	"ESA" in oval, on the left flat. Some rifles will show a script *N* (reading sideways) immediately behind the trigger plate
Principal Dimensions	
Overall length	48 inches
Stock Length	44-3/4 inches; barrel band shoulders 16-1/8 inches apart
Barrel Lengh	33 inches overall, 29-5/8 inches in bore; 3 grooves, 1 turn in 42 inches to the right. Barrels were lined
Muzzle Diameter	0.742 inch
Ramrod Length	32-9/16 inches (nearly flush with the muzzle when stowed)
* The term "blackened" is used to describe the color achieved by case-hardening in oil and water. The oil was floated on water. When the part to be case-hardened was removed from the oven, it was immediately immersed in the oil-water bath. The oil produced the black color and the water hardened the steel.	

Identifying Features

Stock

The Cadet Rifle's stock was 44-3/4 inches long, but the most noticeable difference was the shorter length of pull of 12-1/2 inches, slightly more than 1 inch shorter than the full-size U.S. Model Springfield 1866 Rifle, see Figure 5-2. This shorter stock was used only on this model. It is similar in configuration to the buttstock found on the U.S. (Navy) Model 1867 .50-45 Cadet Rifle (built on the Remington rolling block action, see Chapter 12). Two barrel bands were used with shoulders spaced 16-1/8 inches apart. A screw-in ramrod, with a concealed threaded ramrod stop inletted just above the trigger plate, was used just as on the service rifle. The wood was greatly slimmed down, approximately 0.125 inch narrower in the lock area, requiring the use of noticeably smaller-diameter barrel bands, shorter lock screws, etc.

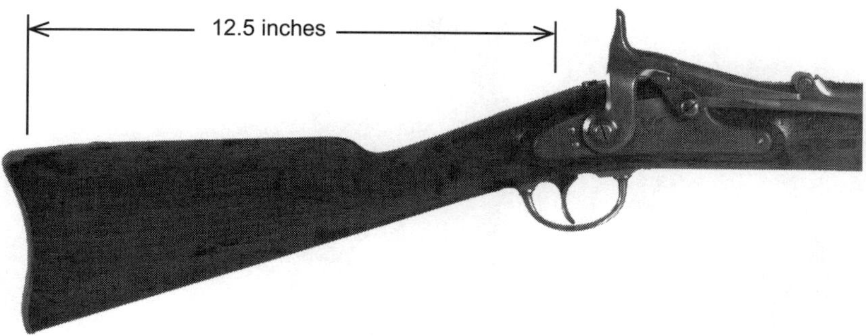

Fig. 5-2. The buttstock was slightly more than one inch shorter than the full-size U.S. Model 1866 Springfield Rifle. The lock plate bevel was also thinner than the full-size rifle lock plate (0.190 inch vs. 0.300 inch). Photograph courtesy of Springfield Armory NHS, SPAR #1129.

The Model 1866 (1867) Cadet Rifle was inspected by Master Armorer Erskine S. Allin. His cartouche, *ESA*, in script inside an oval, was stamped on the left stock flat behind the lock plate screws, see Figure 5-3. Note also the stylized "eagle head" stamped on the left rear of the receiver. The script *N* was used to denote that the rifle had been proof fired, see Figure 5-4.

NOTE: The example shown in Figure 5-3 is from the Springfield Armory Museum collection, SPAR #1129, and does show inspector's marks from its previous incarnation. But examples observed by an authority on this period, Ed Hull, and the author, lack these markings and show only the later "ESA" cartouche in an oval, characteristic of this post–Civil War period. SPAR #1129 may have originally been a prototype made using a cutdown Model 1864 percussion rifle musket stock.

Barrel

The barrel was 33 inches long overall and 29-5/8 inches long through the bore. It was chambered for the .50-70 cartridge but it was intended that a lighter 55-grain loading be used. It was sleeved in the same manner as the 1866 rifles. The rifling was of typical Springfield style, three grooves, right-hand twist, one turn in 42 inches. The barrel ta-

Fig. 5-3. Erskine S. Allin's cartouche will be found on the left side of the U.S. Model 1866 Cadet Rifle stock (arrow A). Note also the "eagle head" marking on the left side, rear of the receiver (B). Photograph courtesy of Springfield Armory NHS, SPAR #1129.

pered to 0.742 inch at the muzzle, to accept the scaled-down Model 1866 Cadet bayonet.

Fig. 5-4. The U.S. Model 1866 Cadet Rifle bears a script "N" firing proof mark. Photograph courtesy of Springfield Armory NHS, SPAR #1129.

Lock

The lock plate was specially manufactured, and was used only on this model, see Figure 5-5. It had the normal beveled edge, but was only 0.190 inch thick, rather than the standard 0.300 inch, and was dated "1867" only. The hammer offset was slightly reduced by 0.110 inch to compensate for the thinner lock plate. Interior parts were of standard musket size and configuration, but the lock plate screws were approximately 0.190 inch shorter, due to the slimmer stock and thinner lock plate.

Fig. 5-5. Two U.S. Model 1866 Cadet Rifle lock plates. Note the date, "1867." The top rifle is from the William Rutter collection and the bottom rifle from the Springfield Armory NHS collection, SPAR #1129.

Breechblock

The breechblock was modified from that used on the parent-rifle by two large semicircular millings, on the bottom, one on each side, to reduce weight; this idea was the forerunner of the large milled "arch" which occurred on the underside of all breechblocks (not just Cadets). This approach began with the 1868 model, and was used up through the first two versions of the 1873 breechblock.

However, on this model—and this model alone—a "rib" of metal, approximately 0.09 inch wide, was left untouched in the center, giv-

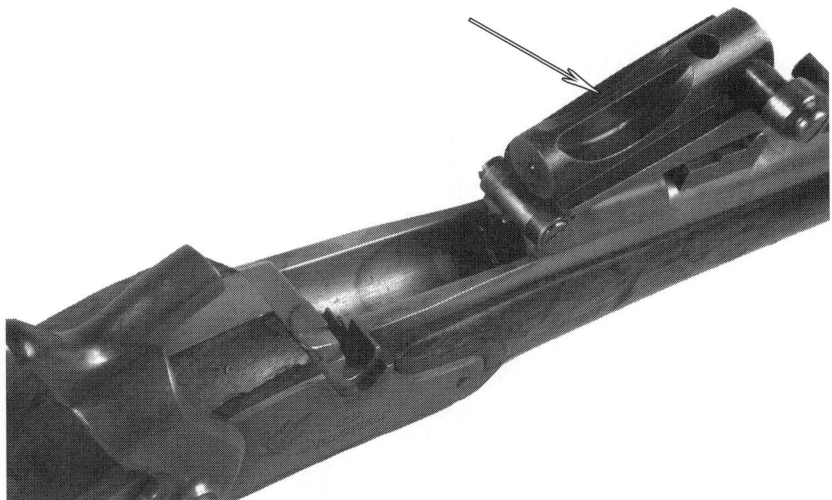

Fig. 5-6. The unique breechblock designed for the U.S. Model 1866 Cadet Rifle was relieved on either side (arrow) to reduce weight, leaving a central rib. Photograph courtesy of Springfield Armory NHS, SPAR #1129.

ing the underside of the block a most unique appearance, see Figure 5-6, arrow. Cadet breechblocks are also known with both styles of cam-shaft attachment, as previously described for the 1866 Rifle. This would seem to indicate use of some leftover, partly machined, early-type 1866 blocks, as the cam change occurred before the first Cadet Rifles were produced.

NOTE: It has recently been discovered that a few very early full-size U.S. Model 1866 .50-70 Springfield Rifles may have also used this breechblock.

Furniture

Many seemingly insignificant parts of this arm are special, and *not* interchangeable with any other Allin-system rifle. In addition to the parts noted above, the trigger guard plate was thinner than standard, and the trigger guard bow was narrower, see Figure 5-7. The two barrel bands are made of thinner material, and are of reduced diameter to fit the slimmer stock and smaller-diameter barrel. Even the band springs are shorter! The nose cap was smaller in all three dimensions, as well as being formed of thinner stock. The butt plate was shorter and narrower. Some of these special parts are interchangeable with the 1867 Navy Cadet model, made at Springfield Armory using the Remington rolling block action, see Chapter 12.

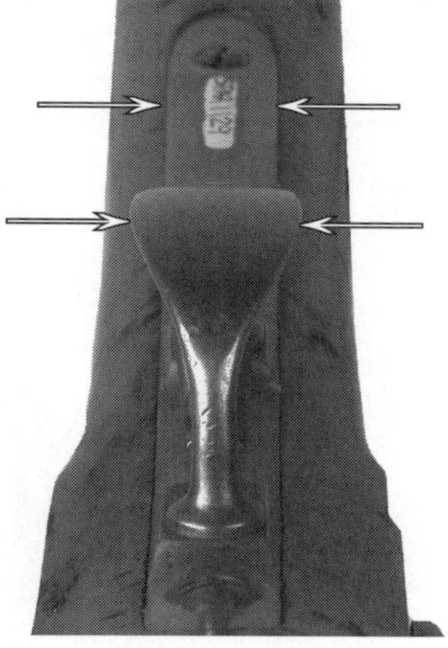

Fig. 5-7. The trigger guard and plate used on the U.S. Model 1866 Cadet Rifle was narrower than on the service rifle. Photograph courtesy of Springfield Armory NHS, SPAR #1129.

Ramrod

Unlike that of the parent U.S. Model 1866 Rifle, this rod extended nearly to the muzzle when stowed, see Figure 5-8. It was 32-9/16 inches long, and had a scaled-down cupped and slotted 7-ring head. The ramrod head design was similar

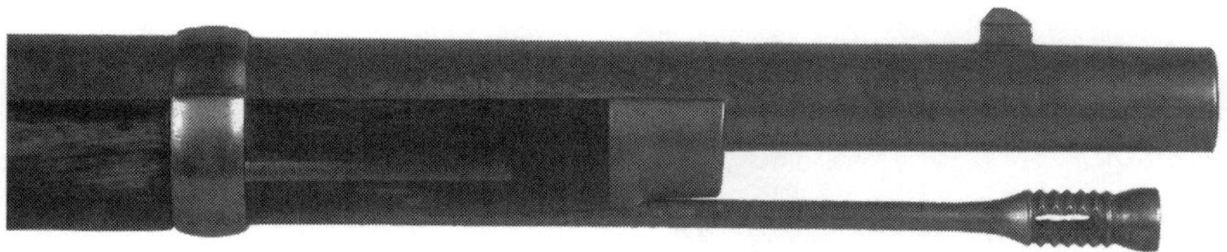

Fig. 5-8. The ramrod used on the U.S. Model 1866 Cadet Rifle was shorter and narrower in diameter than the rifle rod, and when stowed reached just short of the muzzle. Photograph courtesy of Springfield Armory NHS, SPAR #1129.

to the rifle, but smaller, and the shaft was noticeably thinner at approximately 0.203 inch near the head, tapering to 0.191 inch. The end of the ramrod had a #12-26 thread and screwed into a retainer plate above the trigger guard. A similar rod, nearly identical except for being an inch shorter, was used on the 1867 Navy Cadet Rifle.

Chapter 6: The U.S. Model 1868 .50-70 Springfield Rifle and Carbine

Historical Background

While the Model 1866 rifles had proven entirely satisfactory in actual service, it was eventually decided that a separate receiver was desirable, to provide a stronger mounting point for the breechblock, rather than simply continuing to solder and screw a hinge to the musket barrel. It was also realized that the extraction system could stand further improvement. These two basic changes resulted in a new arm, the U.S. Model 1868 .50-70 Rifle, see Figure 6-1. The new model rifle was exactly 4 inches shorter than the Model 1865 .58 Caliber Rifle and the Model 1866 .50-70 Rifle, which had been almost wholly converted from the Model 1861 or Model 1863 series percussion rifle muskets.

This new model rifle can be quickly distinguished by the fact that it has a separate receiver into which the barrel was screwed and only two barrel bands. The Model 1868 .50-70 Rifle would set the general configuration for the service rifle until the end of "trapdoor" production, in late 1893.

A number of unique specimens from this developmental period exist today in the Springfield Armory Museum, and show some of the different ways in which they experimented with fastening the receiver to the barrel. A few even involved wraparound clamps! It is also quite interesting to note that no problems were documented with the lined barrels at

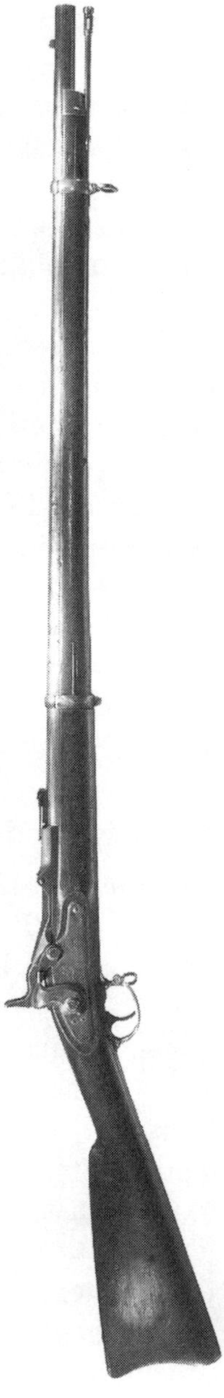

Fig. 6-1. U.S. Model 1868 .50-70 Springfield Rifle, serial #25755.

the time (or since) as used on the two previous rifle models, the U.S. Model 1866 .50-70 Springfield Rifle and the U.S. Model 1866 (1867) .50-55 Springfield Cadet Rifle. In fact, lined barrels continued to be used in early Model 1868 production.

It has been suggested that any U.S. Model 1868 rifle with a lined barrel represents one of the "extra rifles" produced for sale during the Franco-Prussian War. The author disagrees since the Model 1868 Rifle did not become obsolete until after that conflict had ended. Input from other collectors is welcome.

The Model 1868 rifle has been occasionally (and incorrectly in the author's opinion) referred to as the "Third Allin Alteration," even though it was essentially a brand-new arm. This term apparently arose because Congress had forbidden the construction of "new" arms until the Ordnance Department selected a single breechloading system. In any event, the term is not used today.

The Model 1868 rifle did continue to use some surplus musket parts, but, the same could be said of the final "trapdoor" model, the .45-70 Model 1888 rod bayonet rifle! It is simply a matter of deciding where to draw the line.

Quantity Produced

A total of 52,145 rifles were produced, from 1868 through 1870. An additional 1,906 were "assembled from parts" for sale abroad. This number *may* include those rifles with lined barrels.

Unique to this model, the breechblock was actually dated with the year of manufacture. The year "1868" marking is quite rare, as only a very few (possibly less than 100) were made during that first year of production. Several low-two-digit serial-numbered examples are in the Springfield Armory Museum.

NOTE: Records show that just four prototype carbines were also manufactured. However, in the author's opinion, this does not justify recognition of the carbine as an official "U.S. Model 1868 .50-70

Springfield Carbine." It is known that they were not all identical; at least one short-stocked arm with a long-nosed 1868 receiver is still in the Springfield Armory Museum. The Ordnance Department and the Springfield Armory did not produce *any* of the various .50-caliber carbines in anything even approaching sufficient volume to arm the mounted branch. None progressed past the "field trials" stage.

Springfield Armory also produced, in 1869, at the "suggestion" of no less a personage than General William Tecumseh Sherman himself, a very few (perhaps only three) .50-70 caliber "trapdoor" pistols using a lightened form of the long-nosed receiver. This monster did not use scaled-down parts; it was over 18 inches long, and weighed almost 4-1/2 pounds. These pistols were envisioned as being used at close quarters by the cavalry, possibly with multi-ball cartridges. One can only imagine how clumsy the handling and loading of such an arm would have been, to say nothing of the muzzle blast, and recoil! The project never progressed beyond the original testing. One pistol is now at Springfield Armory; one is at Rock Island Arsenal; the other is in the Smithsonian.

Adapted From

While it continued to utilize many small parts from the muzzleloading muskets, this arm, with its separate receiver, and use of (mostly) new, unlined barrels, was considered to be the very first newly made "trapdoor" rifle. This was as opposed to a conversion, since no given specimen had ever previously existed as a complete gun.

Table 6 The U.S. Model 1868 .50-70 Springfield Rifle and Carbine	
Finishes	
Receiver	Blackened*
Breechblock or Bolt	Blackened*
Hammer and/or Lock Plate	Color Case-Hardened
Barrel	National Armory Bright

The .58- and .50-Caliber Rifles

Furniture	National Armory Bright (unless other-wise noted) Rear Sight: Browned (blued)
Stock	Oil-finished American Black Walnut
Markings	
Receiver	Serial number (1-52,300+) at left front, just above wood (should match serial number stamped on the barrel adja-cent)
Breechblock or Bolt	"1868," "1869," or "1870/eagle head/ crossed arrows/U.S."
Hammer and/or Lock Plate	Hammer: knurled in shield pattern Lock Plate: "U.S./SPRINGFIELD" "1863" or "1864" at rear of plate
Barrel	Serial number (1-52,300+) at left rear, just above wood (should match serial number stamped on the receiver adja-cent). Witness marks at right rear, just above wood
Furniture	Butt Plate: "U.S." on tang Bands (2): "U" near upper edge, right side Rear Sight: "2," "3," "5," "7," "9" on right rear face of leaf
Stock	From two to four cartouches with vari-ous initials on left flat (refer to text). "BL" in box behind trigger plate
Principal Dimensions	
Overall Length	52 inches
Stock Length	48-3/4 inches; barrel band shoulders 19-1/8 inches apart
Barrel Length	32-3/4 inches (32-5/8 inches in bore); 3 grooves, 1 turn in 42 inches to the right. A few barrels were lined only in early production; all others were newly manufactured as unlined barrels
Muzzle Diameter	0.775 inch
Ramrod Length	35-5/8 inches (flush with the muzzle when stowed)

* The term "blackened" is used to describe the color achieved by case-hardening in oil and water. The oil was floated on water. When the part to be case-hardened was removed from the oven, it was immediately immersed in the oil-water bath. The oil produced the black color and the water hardened the steel.

Identifying Features

Stock

The stock was newly sized for this model. It had been reduced to 48-3/4 inches in length, including butt plate and tip, with inletting added for the new separate receiver, and the hinge pin retaining arm on the left side. Also new was the change to two barrel bands, the shoulders for which were 19-1/8 inches apart.

At the same time, the mortising for the "horseshoe"-shaped extractor spring as well as for the cartridge retaining spring of the 1866 type was eliminated, although a few stocks are known with "dual" Model 1866 and 1868 inletting. Also, many Model 1868 stocks will be found with inletting for the spoon, even though it was no longer used to retain the ramrod.

Most U.S. Model 1868 rifle stocks were reused from U.S. Model 1863 (no band springs) rifle muskets. The stock will normally show four cartouches in flattened ovals and of slightly differing sizes on the left stock flat. Reading from left to right they should be: 1) "ESA," 2) "SWP" or "FWS" or "TWR," 3) "FG" or "WPT" or "HSH" or "TWR" or "JS," and 4) "ESA." The first two cartouches relate to the conversion of the stock for the Model 1868 rifle; the latter two relate to the stock when originally inspected as a rifle musket. Several other combinations have been seen. The reader is referred to the discussion under "Stock" in Chapter 3 concerning this same issue.

Receiver

A new part was introduced with this model to provide a much stronger mounting point for the hinged breechblock. Where the previous two models had depended on a hinge strap screwed and soldered to a modified barrel, in the Model 1868 the receiver was threaded at the front to accept the new, shorter barrels, see Figure 6-2. It was also threaded at the rear to accept the breech screw. A blind hole, nearly parallel to the bore, in the upper left side housed the new ejector and its spring. The long 1868-style receiver (also used on the U.S. Model 1869 Cadet, see Chapter 7) had one obvious feature which immediately set it apart from the Model 1870, as well as all later .45-caliber versions: the front part ahead of the hinge pin was 23/32 inch (0.72 inch) longer.

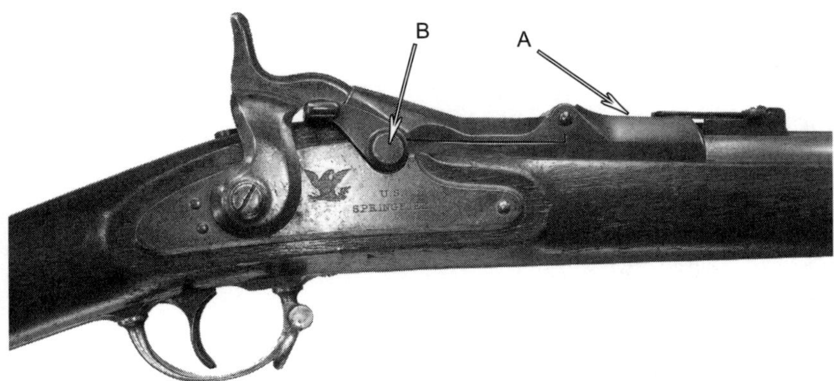

Fig. 6-2. Close-up of the right side of the U.S. Model 1868 action and lock. Note the new separate receiver with the long "nose" (A) that was unique to the U.S. Models of 1868 and 1869. Also note that the thumb latch was no longer screwed to its shaft but was press-fitted (B).

Serial numbering of the Allin-system arms began with this model. The serial number was stamped on the left side of the receiver, just above the stock line and parallel to the bore. It read from muzzle to breech, starting at the barrel joint, see Figure 6-3. A matching number was stamped on the barrel, immediately to the left of the joint. A mismatched barrel and

Fig. 6-3. All U.S. Model 1868 arms should have matching serial numbers, as shown here. The new long-range sight base nearly touches the receiver.

receiver would be undesirable from a collector's standpoint, but are almost never encountered.

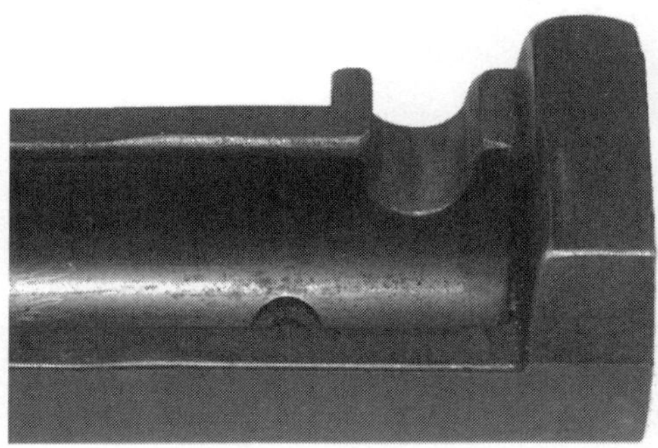

Fig. 6-4. The top rear surface of the receiver extends flat across and does not curve down toward the lock. It is unmarked. The ejector stud is missing in this photo.

The upper rear surface of the receiver (where the serial number was stamped on the .45-caliber arms) was of different shape, see Figure 6-4. It was flat from right to left and did not curve down, toward the lock plate, as do the later, post-1873 arms.

Breechblock

The breechblock was distinctly different in shape from the Model 1866. The comb stopped approximately 1-1/4 inches short of the hinge knuckle, rather than having one edge nearly meeting it. The block was arched on the bottom; and there was a new-style, three-feature marking on top consisting of "year/eagle head & crossed arrows/U.S." Dates can be "1868" (very rare), "1869" (common), or (mostly) "1870," see Figure 6-5 and compare to Figures 3-6 and 8-7.

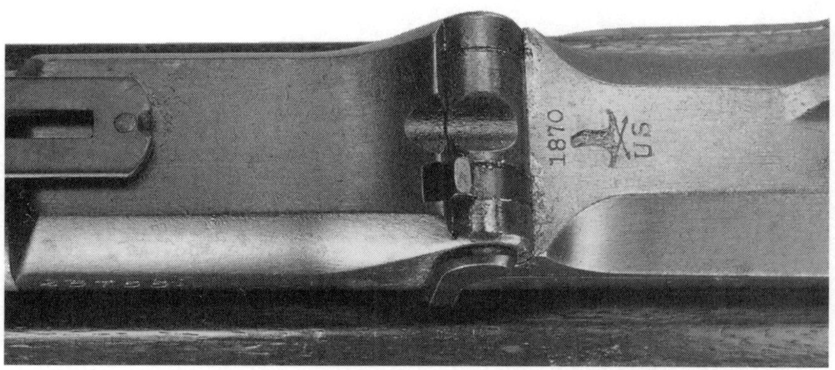

Fig. 6-5. Nearly all U.S. Model 1868 arms are dated either "1869" or "1870" (as shown here). The "1868" date is very rare, occurring on only approximately 100 rifles.

The breechblock, see Figure 6-6, resembles the later U.S. Model 1870 block, but for two points: The arch on the bottom was shorter (1-7/8 inches) and the top flange (arrow A) was considerably thicker (0.368 inch vs. 0.291 inch). When viewed from the side, the top surface of the breechblock was very nearly aligned with the center of the hinge pin (arrow B). This surplus thickness inhibited the opening of the breech, and would subsequently be changed. On the U.S. Model 1870 (and all later) blocks, the top surface aligned near the bottom of the hinge pin; this allowed the breechblock to be opened further.

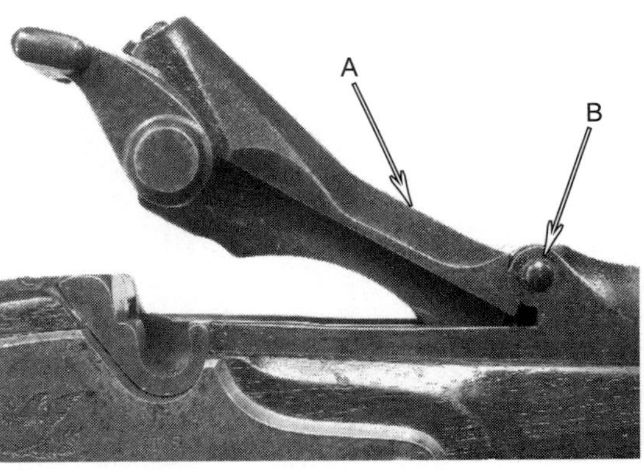

Fig. 6-6. Right-side view of the breechblock, partially opened. Note that the line of top surface (A), which is much thicker on this than on subsequent models, is roughly at the center of the hinge-pin axis (B) and inhibits the degree to which it can be opened.

Fig. 6-7. The new thumb latch was smaller than on previous models and its corners were squared.

The 1868 thumb latch was smaller than in previous models, and had two square outer corners, see Figure 6-7, which differentiated it from the 1870 (front square and rear round) as well as the 1873 and all later models in which both "corners" were noticeably rounded. Compare to Figure 8-9.

The new firing pin was spring loaded, and was retained in the block by means of a screw entering from the underside. The tip of this screw, which could only be accessed after removing the thumb latch, entered a milled slot on the bottom of the firing pin, near its head. The Model 1868 Rifle furnished the prototype for the firing-pin design used on all subsequent Allin arms.

Extractor

The U.S. Model 1868 Rifle marked the first use by Springfield Armory of the familiar blade-type extractor that pivoted around the hinge pin, and was powered by the plunger-style ejector pin with its powerful surrounding spring, see Figure 6-8, arrow. The pin and spring were located in a cavity which was bored out nearly parallel with the bore, in the upper left front of the new separate receiver. This mechanism would remain virtually unchanged for twenty-five years to the end of .45-caliber rifle production, in 1893. A similar extractor had been used on the Miller (commercial) .58-caliber rifle musket conversion, which led to a battle over patent infringement and royalties. The ejector plunger and spring are based on a Berdan design.

Fig. 6-8. A new simplified extractor was developed for the U.S. Model 1868. This same design would be used in all subsequent U.S. rifles and carbines based on the Allin patents.

The new extractor occasionally tore through the soft rim of the early folded-head, inside-primed, copper cases, usually because of dirty ammunition sticking in the chamber; the modern-style solid-rim case introduced in early 1883 essentially eliminated the problem.

NOTE: The copper used to make the cartridge cases at this time was actually an alloy called Bloomfield gilding metal.

Barrel

The U.S. Model 1868 Rifle marked the first use of the new shorter-length barrel which measured 32-5/8 inches through the bore. Some lined barrels, like those used in the manufacture of the U.S. Model 1866, will be encountered in early production, but the vast majority were newly manufactured without a lining. Rifling was the usual three broad lands and grooves, with a twist of one turn in 42 inches. The barrel was uniformly tapered to 0.775 inch at the muzzle to accept the M1855 bayonet.

The barrel and receiver of the U.S. Model 1868 Rifle were aligned by means of witness marks, just above the wood, on the right-hand side. These were two straight lines struck with a chisel-like tool and should always be in precise alignment. If not, the barrel has been removed and replaced. This is especially important to the collector, as later-model arms do not have the matching serial numbers of the Models 1868 and 1869. A single-letter subinspector's initial was stamped above the witness mark on some barrels, usually a "D" or an "X" (see Figure 8-4 in Chapter 8).

Rear Sight

The front-hinged leaf was entirely new, see Figure 6-9. It was longer and was equipped with a rather primitive slide which lacked a friction spring or any means of windage adjustment (arrow A). The sighting notch on the slide was a wide, deep "V" cut (arrow B). The sight base was located approximately 1/16 inch (0.06 inch) in front of the new receiver. It was attached to the barrel as on previous percussion rifle muskets, in a dovetail cut and secured by one spanner-head screw (arrow C). When folded down, the leaf overlapped the receiver by about 1/2 inch (0.5 inch). The placement of the rear sight on the U.S. Model 1868 Rifle barrel was quite different from that of the U.S. Model 1870 Rifle, with which this arm is sometimes confused, see Chapter 8.

Fig. 6-9. The new U.S. Model 1868 "long-range" rear sight was graduated to 900 yards. The leaf (A) and slide (B) are shown folded flat against the barrel in the top view and open somewhat further than in normal use in the bottom view to show the spanner screw (C) which attached it to the barrel. This rear sight lacked a means of adjusting for windage.

Lower Band

The U.S. Model 1868 Rifle was the only rifle in of all the trapdoor variations to utilize the Model 1863 rifle musket lower band, which has a clamping screw, see Figure 6-10. *All* subsequent lower bands were retained by their band spring alone.

Upper Band

The Model 1868 Rifle also marked the first appearance of an upper band which incorporated the sling swivel as part of its original design, see Figure 6-11. On three-band guns, such as the U.S. Models of 1865 and 1866, the sling swivel was found on the middle band.

Fig. 6-10. The U.S. Model 1868 is the only Allin-patent arm which utilizes a clamping lower band.

Ramrod and Ramrod Stop

The U.S. Model 1868 Springfield Rifle also introduced the style of ramrod retention that would last throughout the trapdoor series until the adoption of the U.S. Model 1888 with the ramrod bayonet in 1890. A round-ended ramrod stop (see Figure 6-12), 3.135 inches long by 0.375 inch wide, with a down-

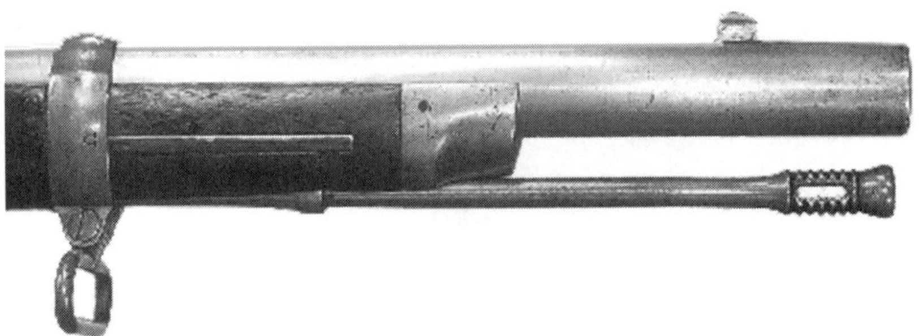

Fig. 6-11. The sling swivel was now mounted forward on the upper band. Also note the single shoulder on the ramrod, which snaps behind the stock-mounted retainer shown in Figure 6-12.

ward projecting lug (arrow A) at either end was mortised into a slot in the barrel (arrow B). This did away with the need to thread the small end of the rod screw into a retaining nut in the stock. It also elimi-

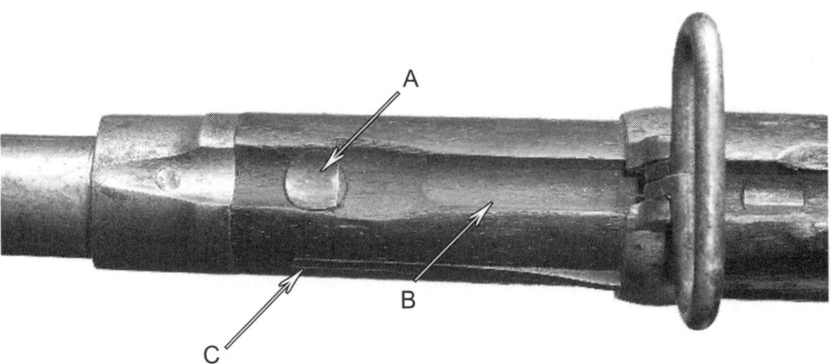

Fig. 6-12. The ramrod stop (arrow A), and its mortise (arrow B) in the forend. The upper band-spring pin (C) passed through the head of the ramrod stop, locking it firmly in position.

nated the spoon and its pin, as well as the retaining nut in the stock's trigger guard mortise. The retaining pin of the upper band spring did double duty, passing through the rear lug of the ramrod stop, holding it in place (arrow C). The front lug of the ramrod stop formed a shoulder which captured an opposing shoulder on the ramrod itself.

The Model 1868 ramrod had a single shoulder, see Figure 6-13 arrow. This would be changed in the later U.S. Model 1870 design, after it was noted that the single square edge of the shoulder battered the muzzle during cleaning. This shoulder battering often formed a ring which is sometimes mistaken for the joint between a barrel and a liner.

Fig. 6-13. This photograph shows the single shoulder of the ramrod which was held in place by the new ramrod stop.

The ramrod itself was 35-5/8 inches long, with a plain lower end without threads or cannelures, and had a seven-ring uncupped head, which was flush with the muzzle when stowed. To remove the rod (which was spring-tempered) it had to be first be pulled away, or sprung "down," from the barrel, to clear the ramrod stop, then pulled straight out.

Chapter 7: The U.S. Model 1869 .50-55 Springfield Cadet Rifle

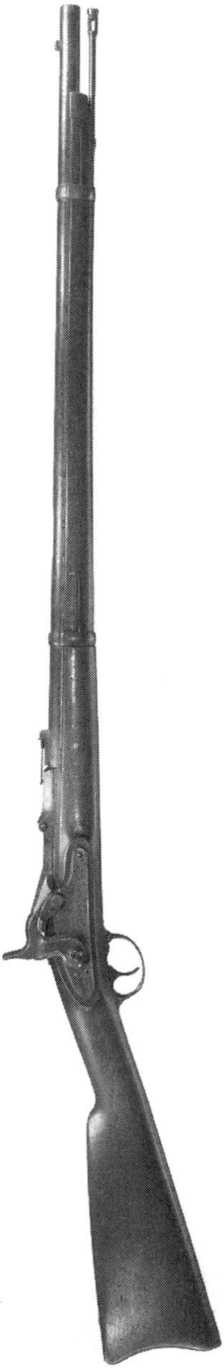

Historical Background

As with the preceding U.S. Model 1866 .50-55 Springfield Cadet Rifle (see Chapter 5), the U.S. Model 1869 .50-55 Springfield Cadet Rifle was an effort to provide an arm of slightly lesser size and weight, better suited to young men who were smaller in stature than their older brethren in the army.

However, this model, even though of reduced length, represents a return to an arm much closer in design to the basic service rifle, as many of the unique but minor, nuisance-like (from the standpoint of interchangeability) differences of the U.S. Model 1866 .50-55 Springfield Cadet Rifle were abandoned. Never again was that near-total miniaturization attempted.

The pattern begun with the 1869 Cadet Rifle laid the groundwork for all succeeding cadet rifles, which totaled nearly 25,000 over the next 25 years or so, see Figure 7-1. The new concept was fairly simple: Shorten the rifles overall by 3 inches, thin the butt, and omit the sling swivels.

NOTE: It is the author's understanding that the U.S. Model 1867 .50-45 Navy Carbine (manufactured by Remington) and the U.S. Model 1867 .50-45 Navy Cadet Rifle (Remington Patents) were the

Fig. 7-1. U.S. Model 1869 Springfield Cadet Rifle, serial number 399. This model was chambered for the .50-55 cartridge. Note that the Cadet Rifle was shorter than the U.S. Model 1868 but the stock was full length in the butt area. It also lacked sling swivels.

only two U.S. model rifles made with the short chamber. The U.S. Model 1869 Springfield Cadet Rifle and the U.S. Model 1870 .50-70 Springfield Carbine fired the .50-55 cartridge, but the cartridge used the full-length .50-70 cartridge case with wads on top of the 55-grain powder charge to take up the excess space.

Quantity Produced

An "official" total of 3,422 U.S. Model 1869 Cadet Rifles were made in 1871 and 1872, including one reported to be manufactured as a "sporting rifle." Serially numbered in the same dual-stamped "barrel-receiver" format as the 1868 Rifle, the 1869 Cadet Rifle had its own separate serial number series, beginning with "1."

NOTE: According to the 1876 Report of the Chief of Ordnance, an additional twenty such arms were produced during the first six months of that year; it has been suggested that they *may* have used the then "short-nosed" 1870/1873-style receiver with the thinner-decked 1870-style breechblock. The author has not seen one but the Springfield Armory Museum has at least two rifles (SPAR #s 5748 and 5778) matching that description. But as with all such museum items, it cannot be taken for granted that these are "production" rifles.

Fig. 7-2. Close-up of the Model 1869 Cadet action; note the similarity to the U.S Model 1868 Rifle. Both were equipped with the long-nosed receiver, except possibly for the twenty cadet rifles manufactured in 1876.

Adapted From

The Model 1869 Cadet Rifle was similar in general overall appearance to its companion U.S. Model 1868 Rifle in that it shared the very distinctive long-nosed receiver, and thick-decked (0.368-inch) breechblock of that arm, see Figure 7-2.

Table 7 The U.S. Model 1869 .50-55 Springfield Cadet Rifle	
Finishes	
Receiver	Blackened*
Breechblock or Bolt	Blackened*
Hammer and/or Lock Plate	Color Case-Hardened
Barrel	National Armory Bright
Furniture	National Armory Bright (unless otherwise noted) Rear Sight: Browned (blued)
Stock	Oil-finished American Black Walnut
Markings	
Receiver	Serial number (1-3,422+) at left front, just above wood (should match serial number stamped on the barrel adjacent)
Breechblock or Bolt	Date of "1870/eagle head/crossed arrows/U.S."
Hammer and/or Lock Plate	Hammer: knurled in shield pattern Lock Plate: "U.S./SPRINGFIELD" "1863" or "1864" at rear of lock plate
Barrel	Serial number (1-3,422+) at left rear, just above wood (should match serial number stamped on the receiver adjacent). Witness marks at right rear, just above wood
Furniture	Butt Plate: "U.S." on tang Bands (2): "U" near upper edge, right side Rear Sight: "2," "3," "5," "7," "9" on right rear face of leaf
Stock	"ESA" in oval, on left flat. "BL" in box behind trigger plate
Principal Dimensions	
Overall Length	49 inches
Stock Length	45-3/4 inches; barrel band shoulders 16-1/8 inches apart
Barrel Length	29-5/8 inches in bore; 3 grooves, 1 turn in 42 inches to the right. Barrels were lined

Muzzle Diameter	0.745 inch (Model 1855 bayonet-type with sleeved socket. The blade was thinned and shortened to 16-1/4 inches)
Ramrod Length	32-5/8 inches (flush with the muzzle when stowed)
* The term "blackened" is used to describe the color achieved by case-hardening in oil and water. The oil was floated on water. When the part to be case-hardened was removed from the oven, it was immediately immersed in the oil-water bath. The oil produced the black color and the water hardened the steel.	

Identifying Features

Stock

While 3 inches shorter overall than the full-sized rifles, the length of pull, which had been reduced in the Model 1866 (or 1867, if you prefer) Cadet Rifle, was restored to the standard 13-1/2 inches but the narrowed butt was retained. The overall length of the Cadet stock was 45-3/4 inches, and the shoulders for the barrel bands were 16-1/8 inches apart. By comparing these measurements with those of the rifle, it can be seen that the reduced length was achieved by shortening the forend between the barrel bands. This was a characteristic that would continue until the end of "Cadet" production in 1893.

Since these shorter stocks were all made new, there should be only one cartouche on the left stock flat on the Model 1869 Cadet Rifle, "ESA" in a small oval, without date; also a small square cartouche, with a "B," an "L," or a "BL" in a box, on bottom of stock, midway between trigger plate and butt. This appears to be an early precursor of the circled "P" indicating that the rifle had been proof fired, found on all stocks beginning with the Model 1873.

A few of the Model 1869 Cadet Rifles (generally those with very low serial numbers) which were overhauled at the Springfield Armory during the time of the "great 1880 turn-in for repair and refurbishment," will show an additional inspector's cartouche, "SWP 1881" (in a rectangle with clipped corners, identical to that found on .45-caliber arms). It was stamped above the original "ESA" cartouche,

Fig. 7-3. The U.S. Model 1869 Cadet Rifle had only one cartouche (ESA). This particular gun was cleaned and repaired at the Springfield Armory in 1881 and so also shows the cartouche (SWP) of Samuel W. Porter. Cartouches are enhanced for clarity.

see Figure 7-3, on the left stock flat. Also, the then-current circled "P" was added just behind the trigger plate, see Figure 7-4.

Barrel

The barrel length was 29-5/8 inches, measured in the bore. The barrel tapered uniformly to 0.745-inch diameter at muzzle to accept the sleeved U.S. Model 1855 bayonet socket. The rear sight was identical in type, markings, and placement on the receiver as that used on the U.S. Model 1868 Rifle, i.e., 1/16 inch (0.06 inch) in front of the receiver (see Chapter 6). The one-piece front sight was similar to that used on the rifle and also served as the mounting stud for the standard-issue Model 1855 bayonet. The barrel and receiver were stamped with matching numbers at their common joint, just above the wood on the left side, see Figure 7-5.

Unlike the U.S. Model 1867 Navy Cadet Remington Rifle (Chapter 12), this arm was chambered for the full-length .50-70 cartridge although the powder charge was limited to 55 grains in order to reduce recoil as much as possible, see Figure 7-6.

Fig. 7-4. Most pre-1873 arms show this boxed "BL" cartouche. This rifle was refurbished and so it also shows the then-current circled "P" indicating that it was again proof fired. Cartouches are enhanced for clarity.

Furniture

The barrel bands used on the U.S. Model 1869 Cadet Rifle were similar to the lighterweight barrel bands used on every Model 1866 Cadet Rifle—slightly thinner than those used on the Model 1868 Rifle. Also, due to the difference in stock taper necessary to reduce the barrel by 3 inches in length, the bands were slightly smaller than, and would not interchange with, those of the Model 1868 Rifle. Full-sized rifle band springs were used to hold the barrel bands in place against recoil.

Fig. 7-5. All U.S. Model 1869 Cadet Rifles should have matching serial numbers on barrel and receiver, below approximately #3,422+, as does the example shown here.

Fig. 7-6. These .50-55 carbine, Benet-primed cartridges were used with the Cadet Rifle, even though full-power .50-70 cartridges could be chambered.

The upper band had no provision for a sling swivel, see Figure 7-7, as cadets did not sling their rifles while marching at the national military academies.

The trigger guard was slimmed down to 0.88 inch wide as against the 0.985-inch-wide trigger guard used on the service rifle. The trigger guard bow did not have a sling swivel, see Figure 7-8. Retained was a narrower (1.47 inches wide vs. 1.68 inches) butt plate, similar to that used with the Model 1866 Cadet Rifle, see Figure 7-9. Both items were slightly smaller than their service rifle counterparts but again the differences were not as extreme as those seen on the U.S. Model 1866 Cadet Rifle.

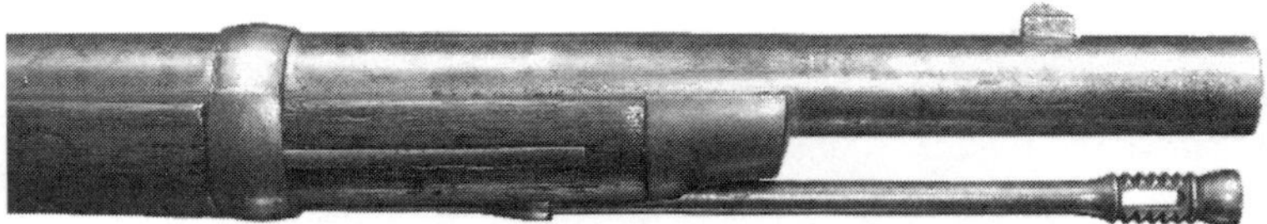

Fig. 7-7. The upper band on pre-1873 Cadet arms did not have a sling swivel as rifle slings were not used at the U.S. Military Academy and most military schools at the time. Note the single-shoulder rod, similar to that developed for the U.S. Model 1868 rifle.

Fig. 7-8. The Cadet trigger guard bow was only 0.88 inch wide (arrows), which was 0.105 inch less than that of the service rifle.

The nose cap was the same size as that of the service rifle.

Ramrod and Ramrod Stop

The M1868-style ramrod with the single-shoulder and seven-ring uncupped head was reduced by 3 inches to 32-5/8 inches in length to fit the shorter barrel. The lower end was not threaded, nor did it have cannelures. The same type of ramrod stop used in the full-sized rifle was mortised into the barrel channel at the upper band to hold the ramrod in place, which was flush with the muzzle when stowed, refer to Figure 7-7.

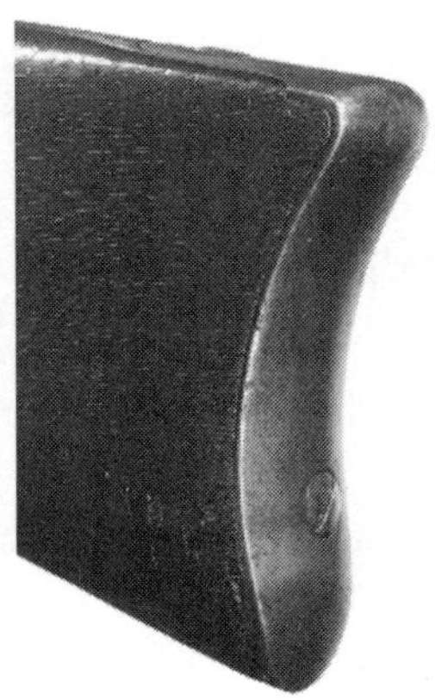

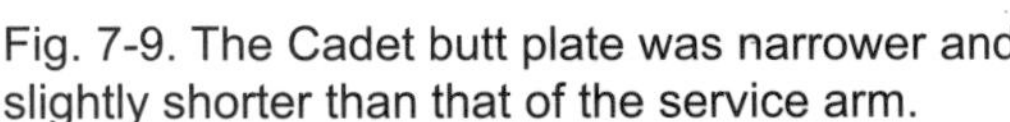

Fig. 7-9. The Cadet butt plate was narrower and slightly shorter than that of the service arm.

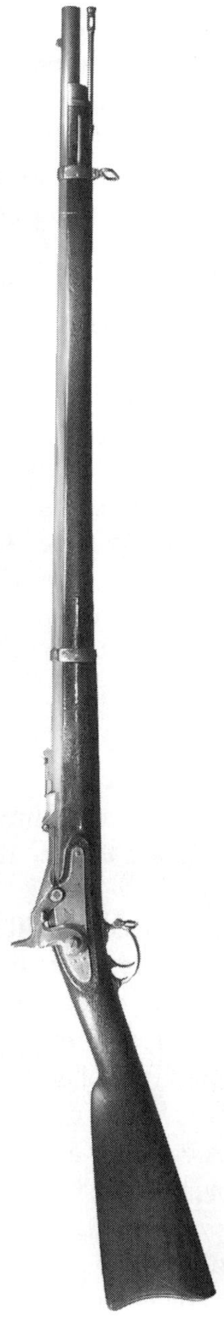

Historical Background

The early 1870s marked a period of change in small arms for the U.S. military. The decision taken in 1865 to proceed with the Allin breechloading design had been reviewed in light of numerous field reports. The Model 1868 had been a vast improvement, but various factions still espoused other breechloading systems, particularly Remington's rolling block design.

In 1869, the "Ordnance Board on Tactics, Small-Arms and Accoutrements" chaired by Major General John Schofield, met at the St. Louis Arsenal, headquarters of the Department of the Missouri. They were charged with selecting a breechloading system for future manufacture. The location was itself a major change; usually such procedures were undertaken in the comfort of a large eastern city, i.e., New York City, Washington, D.C., or Springfield, Massachusetts.

At about the same time, Congress decreed that one system, and one system only, would be used for the manufacture of small arms for the U.S. service. The Ordnance Department tried to make an "end run" around Congress by stepping up the production of the Model 1868, but this plan was discovered, and squelched (see Chapter 6). As a result, appropriations for Springfield Armory were to be severely curtailed until the board had made its final decision.

Fig. 8-1. U.S. Model 1870 Springfield .50-70 Rifle. This example is from early in the second production run.

The .58- and .50-Caliber Rifles

Four arms were recommended by the Schofield Board for a year-long trial in the field, each one to be tested in both rifle and carbine configuration. These arms were: 1) the prototype of the U.S. Model 1870 Springfield Rifle (an Allin design) as shown in Figure 8-1; 2) the Remington rolling block (see Chapter 14); 3) the Sharps (see Chapter 17); and 4) the Ward-Burton (see Chapter 18). Nominal quantities manufactured for the field trials were one thousand rifles and three hundred carbines of each design.

The arm ultimately chosen (and perhaps not unexpectedly) was the Allin design which would become known as the U.S. Model 1870 .50-70 Springfield Rifle. The few changes made from the Model 1868 hardly seem sufficient to justify a new model designation, but it was officially recognized in 1872 as such by the Ordnance Department. While the caliber and cartridge were to be short-lived (both would soon be reduced to .45 caliber by the Terry Board sitting at New York from September 1872 to May 1873), the action of the "Allin system" did serve (surviving several subsequent attempts to substitute other single-shot and primitive magazine arms) until the adoption of the U.S. Model 1892 Krag-Jorgensen nineteen years later.

Quantity Produced

Approximately 12,000 U.S. Model 1870 .50-70 Springfield Rifles were manufactured during the period 1870 through 1873. This total includes the one thousand arms issued for trials.

U.S. Model 1870 Springfield Rifles were apparently *not* serially numbered (at least not to any great extent) although this has been the subject of debate among Springfield scholars. Hopefully some diligent researcher will ultimately uncover the truth. Numerous different marking schemes have been reported for existing rifles.

A very few arms have matching numbered barrels and receivers; these numbers usually occur on the left side, just above the wood, as on the preceding U.S. Model 1868 .50-70 Springfield Rifle, but in an entirely different typeface. Two specimens (numbered 76|76 and 77|77) are known, marked (apparently using the same dies as the Model 1868

Springfield rifle) on the *bottom* of the barrel and receiver, reading from muzzle to breech, see Figure 8-2. The author would be *very* interested in hearing from anyone else who has a U.S. Model 1870 .50-70 Springfield Rifle marked in this particular format, using what certainly appear to be the 1868 dies.

Fig. 8-2. The serial number, "77," struck on the bottom of this early second-production-run Model 1870 barrel and receiver, appears to have been from the same dies used on the U.S. Model 1868 rifle. Number "76|76" has also been reported. This specimen has the first digit (originally a "6") restruck.

The most common marking variation reported so far seems to be a number on the *right* side of the barrel *only*, just ahead of the receiver; however, since many of these numbers are far higher than the quantity of arms known to have been produced, they are generally deemed *not* to be serial numbers.

It is known that one hundred Model 1870 Rifles (presumably included in the 12,000 figure above) were fitted with an early, ten-round version of the Metcalfe cartridge-holding device; they were used in the 1872 trials, and also briefly in regular service. But scant details are available and none of the guns are known to have survived.

Adapted From

As stated, the U.S. Model 1870 .50-70 Springfield evolved from the U.S. Model 1868 .50-70 Springfield. The differences are several, but slight, which is typical of the gradual improvements that occurred in later years with the .45-caliber models.

Table 8 U.S. Model 1870 .50-70 Springfield Rifle	
Finishes	
Receiver	Blackened*
Breechblock or Bolt	Blackened*
Hammer and/or Lock Plate	Color Case-Hardened
Barrel	National Armory Bright
Furniture	National Armory Bright (unless otherwise noted) Rear Sight: Browned (blued)
Stock	Oil-finished American Black Walnut
Markings	
Receiver	None (a very few reported with numbers; possibly serial numbers at the left side, or bottom—refer to text)
Breechblock or Bolt	Date of "1870/eagle head/crossed arrows/U.S." Late-production breechblocks show the word "MODEL" above the year "1870"
Hammer and/or Lock Plate	Hammer: knurled in shield pattern Lock Plate: "U.S./SPRINGFIELD" "1863" or "1864" at rear of plate
Barrel	Believed *none* as originally made. Some have been reported with a number on the top right rear, just above the wood (refer to text). Witness marks at right rear, just above wood (some having a letter adjacent)
Furniture	Butt Plate: "U.S." on tang Bands (2): "U" near upper edge, right side Rear Sight: "2," "3," "5," "7," "9" on right rear face of leaf
Stock	From two to four cartouches, with various initials, on left flat (refer to text). "BL" in box behind trigger plate

Principal Dimensions	
Overall Length	52 inches
Stock Length	48-3/4 inches; barrel band shoulders 19-1/8 inches apart
Barrel Length	32-5/8 inches in bore; 1 turn in 42 inches to the right
Muzzle Diameter	0.775 inch
Ramrod Length	35-3/8 inches (flush with the muzzle when stowed)
* The term "blackened" is used to describe the color achieved by case-hardening in oil and water. The oil was floated on water. When the part to be case-hardened was removed from the oven, it was immediately immersed in the oil-water bath. The oil produced the black color and the water hardened the steel.	

Identifying Features

Stock

It has been generally thought that *all* arms of the 1870 pattern utilized stocks left over from those produced (or salvaged from rifle muskets) for the 1868 model. If so, they should have been inletted for that arm's "long-nosed" receiver. Apparently, the vast majority of Model 1870 rifles did use the earlier Model 1868 stock (which fits perfectly). However, the author has carefully examined a completely genuine specimen, that most definitely possesses original arsenal inletting for the shorter 1870 receiver.

Like the Model 1868, most Model 1870 stocks were originally manufactured for the U.S. Model 1863 Springfield rifle musket which lacked inletting for the band springs. As such, they will usually show four cartouches, in flattened ovals on the left stock flat. Each will differ slightly in size. Reading from left to right, they should be: 1) "ESA" 2) "TWR" 3) "FG" or "WPT" or "HSH" or "TWR" or "JS" and 4) "ESA." The first two inspection cartouches relate to the conversion from rifle musket to rifle; the latter two relate to the stock as a former rifle musket. Other combinations have also been observed. Many stocks will be found with inletting for the ramrod spoon; some will even have the ramrod spoon pin installed.

Receiver

The receiver was similar to the Model 1868 Springfield receiver but 23/32 inch (0.720 inch) shorter, the difference being in that part forward of the hinge pin, see Figure 8-3 (arrow B). The "trapdoor" receiver (arrow B) had now taken the length, and general shape (except for its plain, flat, unnumbered, top rear surface) that would remain constant throughout the production run of the "trapdoor" rifles, ending finally in 1893. A *very* few Model 1870 receivers may bear a serial number stamped on the side or bottom, as discussed above.

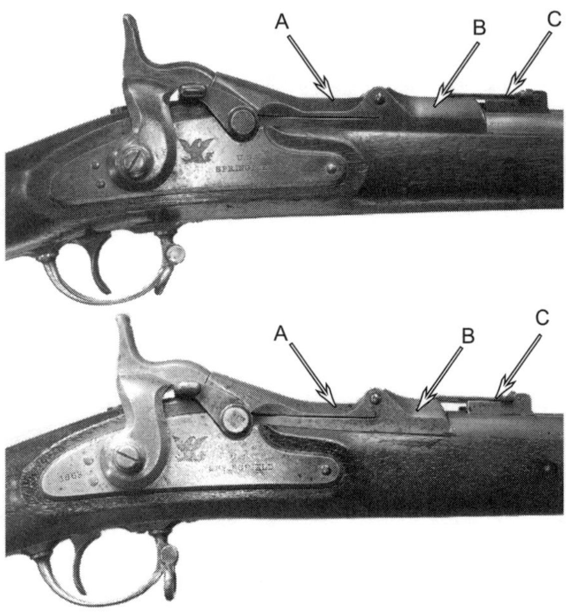

Fig. 8-3. This side view of the U.S. Model 1868 (above) and 1870 (below) rifles shows the difference in the thickness of the breechblocks (A), shorter receiver (B), and difference in the positioning of the rear sights (C).

Barrel

The profile at the rear or breech end of the barrel changed when the shank was shortened ahead of threads to fit the shorter receiver; but the overall length remained unchanged. The length, as measured by the interior bore dimension of 32-5/8 inches, also remained unchanged. Serial numbers were nearly always absent on the Model 1870 barrel (and receiver). The barrel was not proof marked, although though some barrels will show a subinspector's initial adjacent to the witness mark, found just above the wood on the right side, see Figure 8-4. This marking also occurs on some Model 1868 arms, as noted in Chapter 6.

Breechblock

Fig. 8-4. Many U.S. Model 1868 and 1870 rifles will show a subinspector's initial adjacent to the witness mark (usually "D" or "X") on the right side front of the receiver.

The Model 1870 Springfield breechblock was similar to the Model 1868 Springfield breechblock, but for two points, see Figure 8-5. The arch on the bottom was 2-1/4 inches long (arrow A) and the thickness of the top flange was reduced to permit the breechblock to swing further open. From the side, the top flat surface of Model 1870 (and all subsequent) breechblocks will align close to the bottom of the hinge pin (arrow B). On the Model 1868, this same surface is very nearly aligned with the center of the pin.

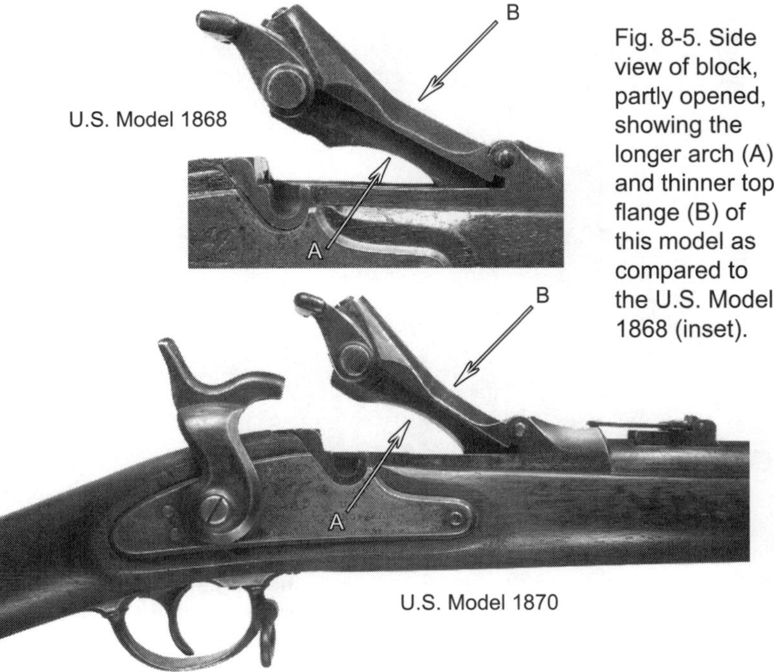

Fig. 8-5. Side view of block, partly opened, showing the longer arch (A) and thinner top flange (B) of this model as compared to the U.S. Model 1868 (inset).

It is unclear when the thinner 1870 breechblock was introduced; it was *possible* that some, or all, of the first 1,000 rifles (*not* the carbines) *may* have been provided with Model 1868 Springfield–style breechblocks.

Model 1870 breechblocks are found with three distinctly different markings:

1) Early breechblocks, including those of the limited run of carbines, are stamped with the identical marking used on the last of the U.S. Model 1868 blocks ("1870/eagle head/crossed arrows/U.S."), see Figure 8-6.

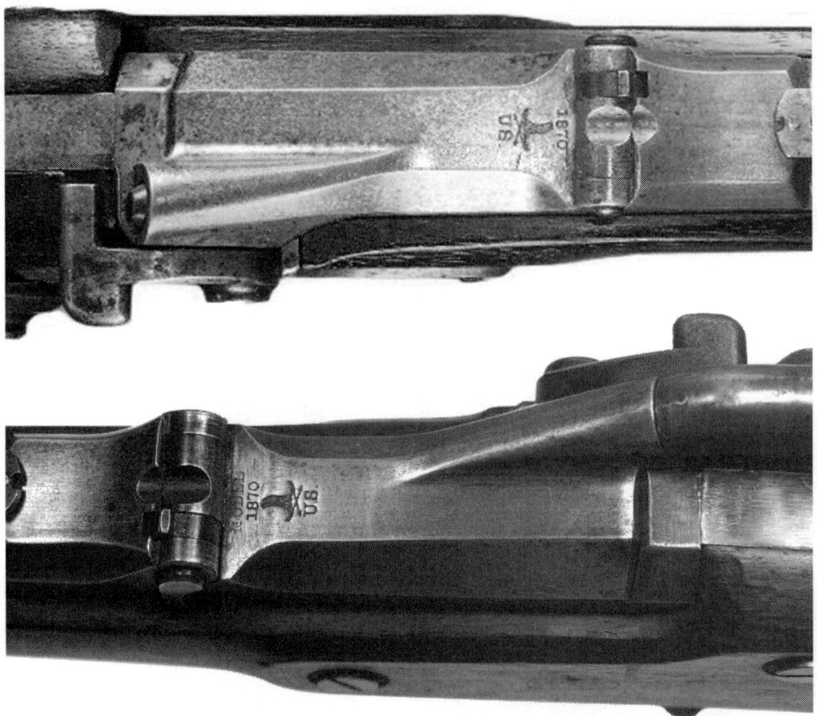

Fig. 8-6. Above: This marking, without word "MODEL," is found on U.S. Model 1870 rifles from the first production run, and much of the second, as well as on all Model 1870 carbines. Below: The later marking with the word "MODEL." Above: Author's collection; Below: Craig Riesch collection.

(2) Later breechblocks, used for rifles only, have the word "MODEL" inserted at the beginning, thus forming a four-feature inscription. Unlike the U.S. Model 1868 breechblocks, which were stamped with the actual year of manufacture, all dated U.S. Model 1870–type breechblocks were marked with the model year, "1870," regardless of the year (1871–1873) they were produced, see Figure 8-7. The marking thus reads: "MODEL/1870/eagle head/crossed arrows/U.S."

Fig. 8-7. This type of marking is only found on U.S. Model 1870 rifles from the later part of the second production run, and never on carbines.

(3) The third variation, and very seldom encountered, has a small eagle (facing left, and of distinctly different form from that used on any other Springfield Armory breechblock) over a small "U.S."; these are believed to have been provided as replacement parts only, as they are undated, see Figure 8-8. It is possible that Springfield Armory did not manufacture this item, although examples viewed were perfectly machined, and clearly quite old. An "1868"-style specimen of this breechblock has also been reported. Again, the author would be glad to receive information on any further examples.

Thumb Latch

Only the rear corner of the thumb latch was rounded off while the front retained the nearly square shape used on the Model 1868 thumb latch, see Figure 8-9. It should be recalled that the final versions of the thumb latch used on .45-caliber arms have both front *and* rear corners rounded.

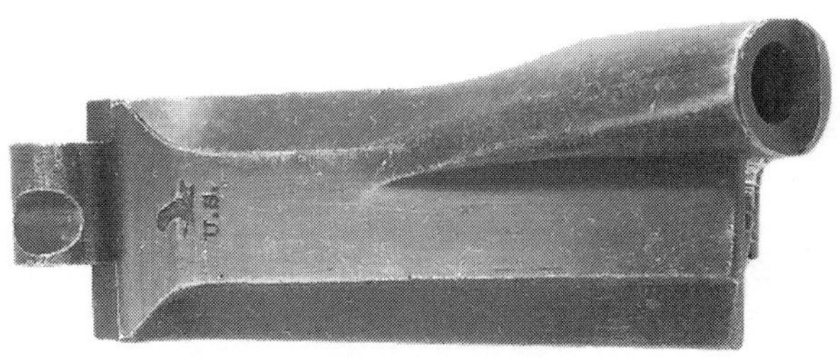

Fig. 8-8. This is clearly a U.S. Model 1870 breechblock, but exact origin and usage is unclear. It is seldom encountered.

Rear Sight

The sight placement on the U.S. Model 1870 Springfield service rifle was noticeably different from that of the U.S. Model 1868 Springfield rifle, with which this arm is frequently confused. The sight base on the service rifle was attached to the barrel (as on previous models) with a dovetail cut and single spanner-head screw 7/16 inch (0.44 inch) in front of the shorter receiver, refer to Figure 8-3, arrow C. Input from collectors having U.S. Model 1870 .50-70 Springfield Rifles with the rear sight placed just 1/16 inch (0.06 inch) ahead of the new short receiver will be gratefully received!

Fig. 8-9. On the U.S. Model 1870 rifles, the rear edge only of the thumb latch is rounded—a transition between the "all-square" thumb latch found on the U.S. Models of 1868 and 1869, and the "all-rounded" 1873 and later models.

When folded down, the leaf just touched the receiver. The rear sight had a front-hinged leaf style, nearly identical to that used on the Model 1868 rifle. The sole change to the Model 1870 rear sight was made to the sighting notch on the slide. Rather than the previous wide, deep

"V" cut, the new, smaller sighting notch was rectangular in shape, with a tiny shallow cut in the center, see Figure 8-10 (arrows). It was a much "finer" but a more difficult sight notch to use.

Fig. 8-10. View of the U.S. Model 1870 rear sight notch. The U.S. Model 1868 rear sight had a plain "V." This new rear sight notch was cut down slightly at sides (arrows) to provide a finer sight.

NOTE: Early production (the first one thousand arms issued for field trials only) Model 1870 rifles have their rear sights located only about 1/16 inch (0.06 inch) ahead of the receiver (the same placement as is found on the U.S. Model 1870 Springfield carbine manufactured at the same time, see Chapter 9). This is also the same relationship between sight and receiver as was found on the U.S. Model 1868 Springfield.

Apparently, the original intent had been to copy that relationship, but for later production it was deemed advisable to move the sight further ahead of the chamber, to avoid weakening the barrel.

One might expect that, when the sight was moved forward for the second production run, it would be returned to the same location, 3-1/8 inches below the lower band, as it had been on the U.S. Model 1868 rifle, but it was not. The maximum adjustment, if the sight leaf was still to rest on the new short receiver, was 3-7/16 inches below the lower band. Since this placed the sight screw well ahead of the chamber, all requirements were satisfied.

Ramrod and Ramrod Stop

The first production run of 1,000 rifles, and (from numerous observed examples) many of the second run as well, retained the single-shouldered ramrod, and rounded ramrod stop of the Model 1868, see Figure 8-11.

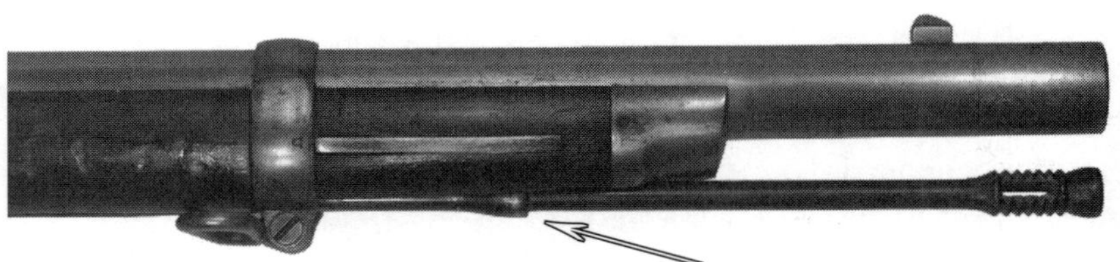

Fig. 8-11. The single-shouldered rod (arrow) survived well into the second production run of the Model 1870 Rifle. Craig Riesch collection.

In late 1871, or early 1872, the ramrod was made with two opposing tapered swells with a notch or groove between for the ramrod stop. This was done to address the complaint that the muzzle was being damaged during routine cleaning, see Figure 8-12, arrow. Naturally, this required that the ramrod stop be reduced to accept the new rod profile, see Figure 8-13. The ramrod was flush with the muzzle when stowed.

The ramrod stops were stamped with the patent date and were used even with the U.S. Model 1873 .45-70 Springfield Rifle until August 1879, see Figure 8-14.

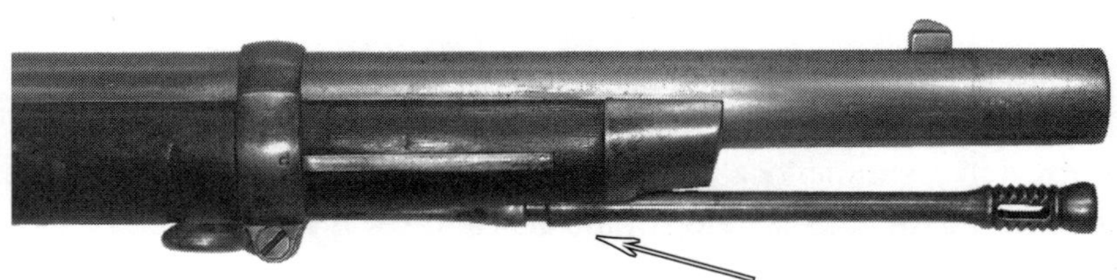

Fig. 8-12. Only late 1871, and 1872 production U.S. Model 1870 rifles were equipped with the double-shouldered (arrow) ramrod. Craig Riesch collection.

Fig. 8-13. The new ramrod stop profile developed for the double-shouldered rod was adopted during the later part of the second production run.

Fig. 8-14. These early double-shouldered ramrod stops bear the patent date. They were also used in the U.S. Model 1873 .45-70 Springfield rifles until August 1879.

Chapter 9: The U.S. Model 1870 .50-70 Springfield Carbine

Historical Background

Congress had stipulated that there would be but one action design throughout the U.S. Army for the shoulder arm, varying only in barrel length and stock configuration to suit the branch of service, as noted in Chapters 6 and 8. Therefore, each of the four rifle designs, ultimately recommended for field trials by the "Ordnance Board on Tactics, Small-Arms and Accoutrements," meeting in St. Louis in 1869–1870, included a carbine for use by mounted troops. The arms finally recommended by the Chief of Ordnance for manufacture as a carbine were a modified Allin system arm referred to in this text as the U.S. Model 1870 .50-70 Springfield Carbine, the Sharps, the Remington, and, almost as a postscript, the Ward-Burton. It should be noted that the Model 1870 Springfield Carbine was not "officially" adopted as such until after the trials were concluded.

Some U.S. Model 1870 Carbines, see Figure 9-1, and some U.S. Model Ward-Burton carbines (see Chapter 18), issued for trial, were carried by Custer's 7th Cavalry on the Yellowstone Expedition of 1873. Many .50-caliber cartridge cases have been found on the route taken by the troops; some show the angled firing pin strike from a Model 1870 Carbine (Allin system),

Fig. 9-1. U.S. Model 1870 .50-70 Springfield Carbine with a 22-inch barrel. Only 361 were made for field trial.

while others show the centered strike from the bolt-action Ward-Burton.

Carbine ammunition used the full-length service copper case (in many of its variants, see Chapter 21: Ammunition for the .58- and .50-Caliber Springfield Arms) with a 55-grain charge and pasteboard filler wads. Overall length of this round was approximately 2.165 inches. As a visual identification, the bullet was much rounder, and deeper set (by approximately 0.08 inch) on the carbine loadings, see Figure 9-2. This enabled the lighter loads to be visually identified when out of their boxes, as the cases were not headstamped.

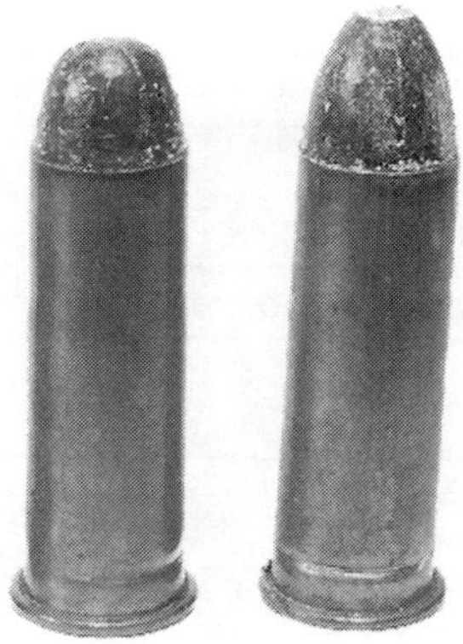

Fig. 9-2. A Benet-primed .50-55 Carbine round, showing the shorter, rounded, bullet (left). A .50-70 rifle round is shown for comparison (right). Except for one run of rifle ammunition in 1882 (R F 9 82), these cartridges were not headstamped.

Quantity Produced

A total of 341 U.S. Model 1870 Carbines were initially made for field trial in early 1871, with another 20 produced during the first part of 1872. These last few were supposedly shipped to San Antonio Arsenal, either as replacements, or for sale to officers. Conflicting records exist, and both versions have their flaws; in any event, no specimen is known to the author which can definitely be proven to have come from that later group.

Adapted From

The Model 1870 Carbine was basically a short version of the 1870 Rifle. Other than the obvious differences in barrel length, and stock configuration, the only other significant design variation had to do with the location of the rear sight, see Figure 9-3. Only on the carbines and the first 1,000 rifles will you find the rear sight nearly touching the receiver. This arm set the basic style, and familiar profile, of the Allin-system "trapdoor" *carbine*, for the next twenty-plus years.

Table 9 U.S. Model 1870 .50-70 Springfield Carbine	
Finishes	
Receiver	Blackened*
Breechblock or Bolt	Blackened*
Hammer and/or Lock Plate	Color Case-Hardened
Barrel	National Armory Bright (a few reported browned/blued)
Furniture	National Armory Bright (unless otherwise noted) Rear Sight: Browned (blued)
Stock	Oil-finished American Black Walnut
Markings	
Receiver	None
Breechblock or Bolt	Date of "1870/eagle head/crossed arrows/U.S." Must *not* show the word "MODEL" above the year "1870"
Hammer and/or Lock Plate	Hammer: knurled in shield pattern Lock Plate: "U.S./SPRINGFIELD" "1865" at rear of plate
Barrel	Witness marks at right rear, just above wood
Furniture	Butt Plate: "U.S." on tang Band (1): "U" near upper edge, right side Rear Sight: "2," "3," "5," "7" only on right rear face of leaf
Stock	"ESA" in oval, on left flat. No date "BL" in box behind trigger plate
Principal Dimensions	
Overall Length	41-5/16 inches
Stock Length	29-3/4 inches
Barrel Length	22 inches in bore; 1 turn in 42 inches to the right
Muzzle Diameter	0.789 inch

* The term “blackened” is used to describe the color achieved by case-hardening in oil and water. The oil was floated on water. When the part to be case-hardened was removed from the oven, it was immediately immersed in the oil-water bath. The oil produced the black color and the water hardened the steel.

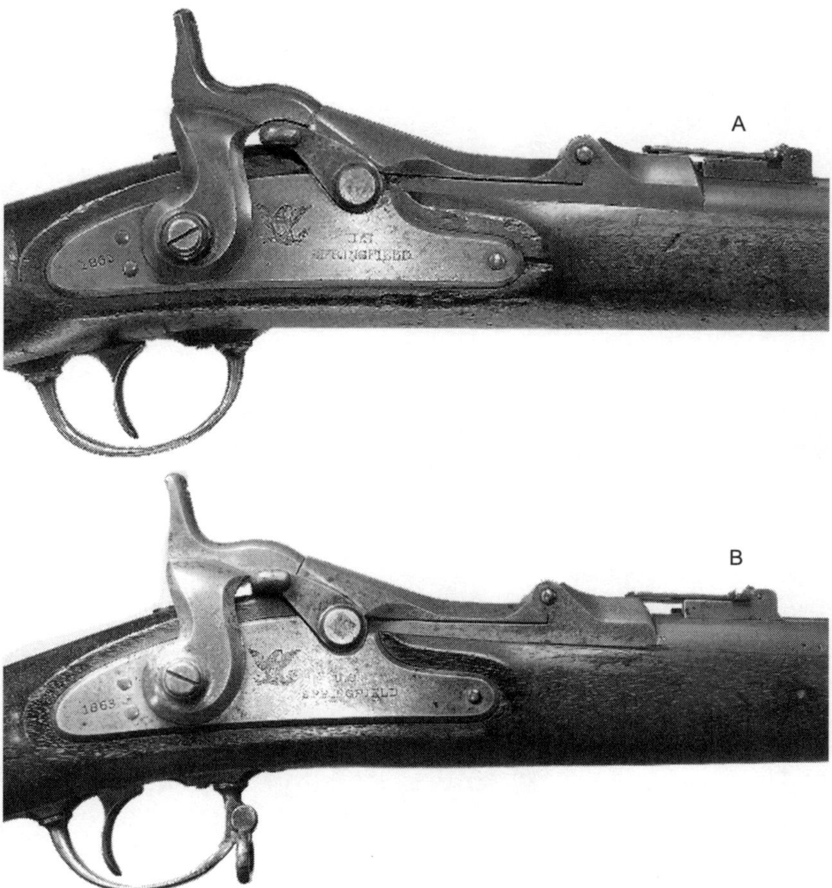

Fig. 9-3. The rear sight on the U.S. Model 1870 Springfield Carbine (top) and the first 1,000 U.S. Model 1870 Rifles was placed directly against the receiver nose (A). Later U.S. Model 1870 Rifle rear sights (bottom) were located forward of the receiver nose (B).

Identifying Features

The carbine differs from the U.S. Model 1870 rifle as follows: 1) the overall length of the arm was reduced to 41-5/16 inches, 2) the forearm was neatly rounded (see Figure 9-4), which eliminated the metal nose cap used on the rifle, 3) the forearm ended 3-1/2 inches ahead of the barrel band, 4) the forearm was not drilled for a ramrod, nor should

Fig. 9-4. The U.S. Model 1870 Carbine, the first manufactured at the Springfield Armory in more than prototype quantities since the U.S. Model 1855, established the standard stock nose profile for the next twenty-five years.

it ever show *any* type of filling where one had been present, 5) the side directly opposite the lock plate was flat rather than rounded as in the later .45-70 carbine stocks, and 6) it was inletted opposite the lock for a carbine sling bar which had flat-faced, oval-shaped bases unlike the later 1873 types which were ground to match the curve of the stock, see Figure 9-5. The lock plate screws ran through the stock and their heads were seated in counterbores in

Fig. 9-5. Note that the bases for the swivel ring bar were finished bright and were flat (arrows), to complement the shape of the stock itself. Craig Riesch collection.

the carbine sling bar bases, rather than the round escutcheons found on rifles.

Stock

The carbine stock was 29-3/4 inches long overall and had a solid butt with no trap for tools nor did the butt plate have a trapdoor. It did not have a ramrod channel and the U.S. Model 1863 rifle-musket-type butt plate was used. Since these stocks were *always* made new, all had only a single cartouche at the rear of the left stock flat, "ESA," in a flattened oval, see Figure 9-6, arrow A. The proof mark, "BL" was stamped behind the trigger plate. Arms carried on the carbine sling for any length of time will always show a pronounced depression worn into the upper right corner of the stock flat. This wear mark is always much more distinct on U.S. Model 1870 carbines than on later ones, because of the raised edge of the flat, refer to Figure 9-6, arrow B.

Fig. 9-6. U.S. Model 1870 carbines will show only one cartouche ("ESA"), see arrow A. If the carbine was issued to a cavalry unit, it will probably show wear by the sling's swivel (arrow B) due to the sharp edge of the wood. This specimen saw some saddle-time!

NOTE: As with *any* Springfield carbine of the period, the buyer needs to be *very* much on the alert for a "carbine" made with an altered rifle stock! Look very carefully at the stock nose for evidence that a ramrod channel was filled in. Move the barrel band forward and examine the step between the forearm and nose to make certain that a new nose (without evidence of a ramrod channel) was not fashioned and glued in place. Also check the forward end of the barrel channel for evidence of wood filler. There are no records or other data to suggest that Springfield Armory ever filled in ramrod channels on rifle stocks and then refashioned them into carbine stocks. Examine the inletting for the carbine sling bar as well. The inletting for the sling bar bases should be crisp with sharp edges.

Barrel

The barrel length through the bore was 22 inches and the diameter at muzzle was 0.789 inch. This was not the same taper as a rifle barrel. Proper crowning was rounded, like the rifle. These carbines saw hard use; those with barrels damaged at the muzzle were usually recrowned, which left a relatively sharp outer edge that sloped towards the inside edge of the bore, see Figure 9-7. This is a difficult feature to see, and the observer will rarely have the opportunity to compare a recrowned barrel to an unmodified one.

Fig. 9-7. The U.S. Model 1870 rifle crown, left; right, recrowned Model 1870 carbine muzzle to repair damage.

The rear sight dovetail slot was cut 1/16 inch (0.06 inch) in front of the receiver, a location shared with the first 1,000 Model 1870 rifles.

NOTE: Measure the barrel diameter at the muzzle of a U.S. Model 1870 Springfield Carbine carefully. It should be 0.789 inch in diameter. If a rifle barrel 32-5/8 inches long was cut off to make a spurious 22-inch-long carbine barrel, its diameter would be 0.825 inch. On the other hand, if the U.S. M1869 Springfield Cadet barrel, 29-5/8 inches long, was cut at 22 inches, its muzzle diameter would be 0.785 inch. This is very close, and it is likely that barrel blanks intended for Model 1869 Cadet rifle production *were* used for the small run of Model 1870 Carbines. The barrel band used on the Model 1869 Cadet rifle was also used on the Model 1870 Carbine.

Breechblock

All Model 1870 Carbine breechblocks were made with the arch that

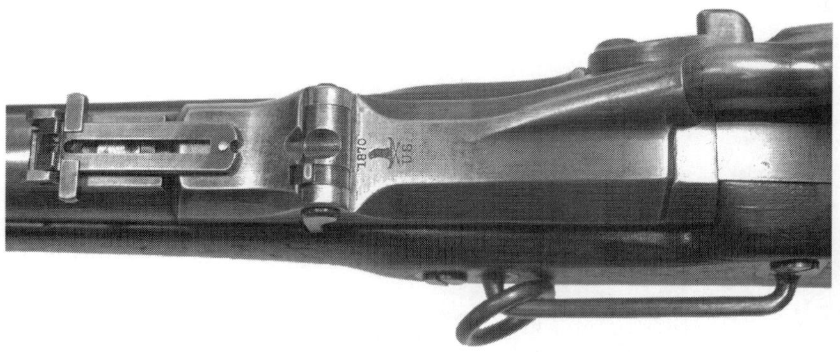

Fig. 9-8. The breechblock marking on all U.S. Model 1870 carbines will read "1870/eagle head & arrows/U.S." The word "MODEL" was not used. Craig Riesch collection.

was 2-1/4 inches long. They were all marked, "1870/eagle head/crossed arrows/U.S.," see Figure 9-8. The word "MODEL," as found on later rifle breechblocks, should *not* appear on the carbine.

Front Sight

The front sight design used on the U.S. Model 1870 Springfield carbine was the first general use by Springfield Armory of a separate, pinned-in replaceable blade, see Figure 9-9, arrow. On the U.S. Model 1870 carbine, this blade was made of brass. Interestingly, the sight blade on the contemporary Ward-Burton carbine (see Chapter 18), also made at Springfield Armory, was steel, as were all subsequent front sight blades. Evidently, the much softer brass blade did not hold up well in service.

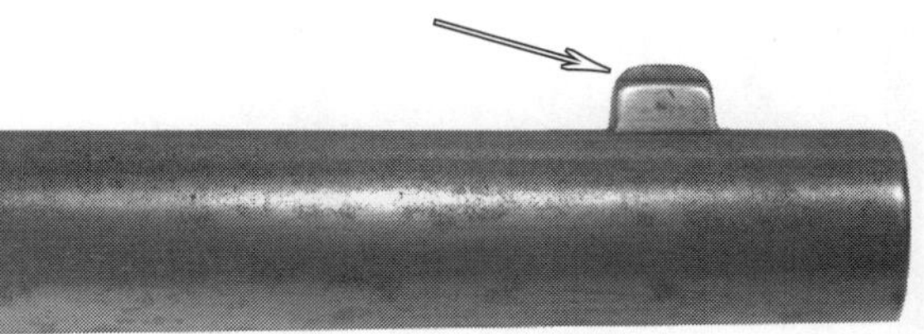

Fig. 9-9. The two-piece front sight, which allowed the replacement of the sight blade (arrow) began with the U.S. Model 1870 carbine. The U.S. Model 1870 carbine front sight was made of brass although the front blade on the Ward-Burton carbine, made at almost the same time, was steel.

Rear Sight

The same rear sight base developed for the U.S Model 1870 Springfield *rifle* was also used on the U.S. Model 1870 Springfield *carbine* but the leaf was graduated differently. Yardage markings were "2," "3," "5," and "7." The "9" [hundred] yard marking found on the rifle slide was omitted. The slide was also considerably smaller, at just 0.169 inch top-to-bottom, than the rifle sight leaf. The distance between the "5" and the "7" on the carbine sight was 0.272 inch, as opposed to 0.408 inch on the rifle sight leaf. This special sight was shared with the three other 1871–1872 trial carbines (Remington, Sharps, and the Ward-Burton), see Figure 9-10.

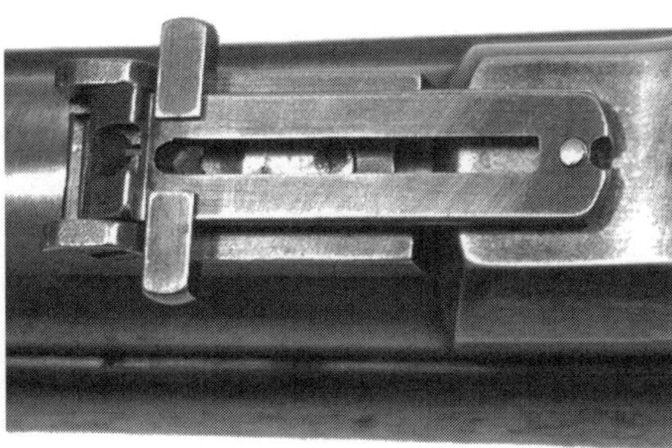

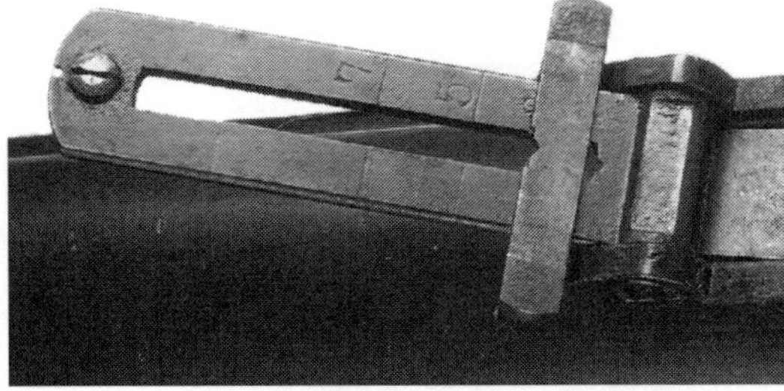

Fig. 9-10. The Model 1870 carbine rear sight leaf, and elevator bar folded flat (above). The carbine leaf is shorter than rifle leaf and graduations are marked only to 700 yards (below) rather than 900 yards.

NOTE: the measurements for the distance between the "5" and "7" on the sight leaves are based on direct observation by the author. Slight variations in measurements can be expected due to the method of measurement used.

Carbine Sling Bar and Ring

The Model 1870 carbine sling bar was similar in appearance to that used on the later .45-caliber carbines, except for two features: 1) the bar and its oval bases were finished bright, and 2) the bases themselves were perfectly flat on their exposed surfaces, since the stocks still had the distinctive flat opposite the lock plate. Beware of any specimen with a blued bar, especially one having the rounded-over bases which did not come into use until the U.S. Model 1873 carbine entered production.

The ring had an outside diameter of 1.061 inches and was formed from iron wire 0.175 inch in diameter. The stock used to form the bar itself was 0.185 inch in diameter. All of these dimensions are exactly the same as found on early Model 1873 Springfield .45-70 carbines. Model 1870 carbine rings and bars can easily be differentiated from the far-more common post-1878 carbine sling bars and rings which were made of much heavier material. The latest (post-1878) bars were made of iron wire 0.222 inch in diameter. The wire used in the post-1878 rings averaged 0.182 inch in diameter, and the finished rings varied from 1.085 to 1.170 inches in diameter.

Fig. 9-11. The trigger guard used on the U.S. Model 1870 Springfield carbine (below) was nearly 1/8 inch (0.13 inch) narrower than that used on the U.S. Model 1870 Springfield rifle (above).

Trigger Guard Bow

The trigger guard bow was narrower (0.89 inch wide) than the rifle trigger guard bow and was not fitted with a sling swivel, see Figure 9-11. It was basically identical to the guard bow of an 1869 Cadet Rifle (see Chapter 7), and quite similar to that used on the Ward-Burton carbine (see Chapter 18).

Barrel Band

The U.S. Model 1870 carbine used a single barrel band. It was made of noticeably thinner material (approximately 0.076 inch) than that of the Model 1870 rifle barrel band, which was 0.090 inch thick. It was marked "U" with the open end pointing toward the muzzle. It was the same rear barrel band that was used on the Model 1869 Cadet rifle, see Figure 9-12.

Fig. 9-12. The barrel band used on the U.S. Model 1870 Springfield .50-70 carbine (left) was thinner than the rear barrel band used on the Model 1873 Springfield .45-70 rifle or the band used on the Model 1877 Springfield .45-70 carbine (right).

Butt Plate

The butt plate used on the Model 1870 carbine was the same butt plate used on the U.S. Model 1870 rifle. It was stamped "U.S." and taken from the large stock of rifle musket parts the Springfield Armory had on hand. It did not have a butt trap for tools.

Ramrod

A ramrod was not used with the U.S. Model 1870 Springfield carbine, nor would be with any carbine until the three-piece jointed ramrod was adopted for the U.S. Model 1877 .45-70 Springfield carbine.

Part II

Non-U.S. Ordnance Department Designs Fabricated or Assembled at the Springfield Armory

1865 to 1871

Chapter 10: The U.S. Model 1865 .50-70 Springfield-Joslyn Rifle

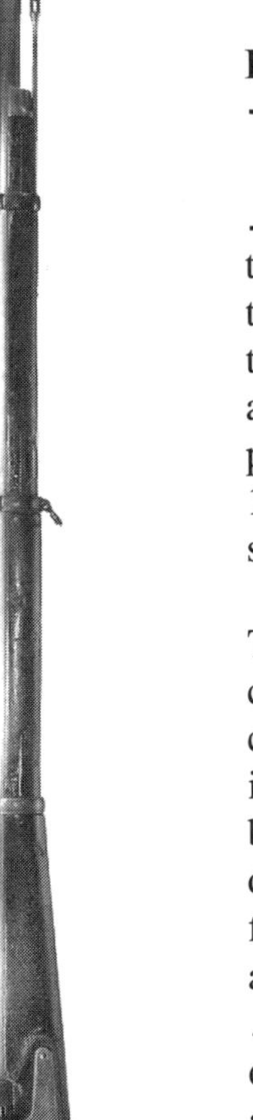

Historical Background

For many years this arm, was thought to be a "conversion" of the rifle musket model, but data uncovered in 1972 made it clear that the rifle was actually newly made. In fact, it was the first arm, expressly designed for metallic cartridges, to be manufactured *in quantity* (the Morse alteration of 1858–59 saw less than 60 guns completed) at Springfield, antedating the U.S. Model 1865 .58 Caliber Rifle (First Allin Alteration) by several months, see Figure 10-1.

These arms were originally chambered for a special Joslyn .50-60 caliber rimfire cartridge, but common .56-.50 Spencer ammunition, although in a shorter cartridge case, fit perfectly, and could be used as well. A second variation of the rifle described here was created in 1871, when Springfield modified the breechblocks and rechambered a number of these guns to accept the then-current .50-70 centerfire cartridge. This was undoubtedly done in order to sell more arms abroad. There was a great effort in the United States to sell arms to France during the period, which has resulted in no small amount of confusion among present-day collectors when trying to classify certain "almost correct" arms of the 1866–1870 era.

In any event, it is known that many of the original rimfire specimens were exported to Central

Fig. 10-1. U.S. Model 1865 .50-70 Joslyn rifle (patent breech), "Type II," centerfire, mixed numbers; note the use of three barrel bands.

America while the centerfire conversions were supposedly all sold to France, almost immediately after being rechambered. However, the Franco-Prussian War ended abruptly; most of the rifles that reached France were ultimately converted to sporting rifles and sold in the African trade. Only a limited number of Springfield-Joslyn rifles remained in, or were returned to, this country, making them quite scarce for today's collector, particularly the modified centerfire variant.

Quantity Produced

A total of 3,007 of the .50-caliber rimfire version were originally assembled at Springfield, early in 1865. It was just barely possible that some of them could have been issued to Union troops before the war ended in April of that year; but, if so, no record has yet been found of their use in combat after the arms were issued on Long Island, N.Y. Of this total, 1,600 rifles were eventually converted to centerfire, at the Armory in 1871, as noted above. The estimated cost of this work, from S. W. Porter's memo of December 19, 1870, was just 35 cents per rifle!

Adapted From

Except for markings, the action was, for all intents and purposes, identical to that used on the Model 1864 Joslyn breechloading carbine, of which 5,000 had been purchased by the Federal government, and used by Union cavalry during the war. Joslyn supplied the complete breech systems for these new rifles, see Figure 10-2, and Springfield produced the rest of the rifle.

Table 10 U.S. Model 1865 Joslyn Rifle	
Finishes	
Receiver	Blackened*
Breechblock or Bolt	Blackened*
Hammer and/or Lock Plate	Color Case-Hardened
Barrel	National Armory Bright

Furniture	National Armory Bright (unless otherwise noted) Rear Sight: Browned (blued)
Stock	Oil-finished American Black Walnut
Markings	
Receiver	Tang: Serial number, from "M/1" to "M/3025+." On original rimfire rifles, numbers should all match
Breechblock or Bolt	Top, and rear face (concealed) of extractor plate: Serial numbers from "M/1" to "M/3025+." Rear: Joslyn patent dates
Hammer and/or Lock Plate	Hammer: knurled in shield pattern Lock Plate: Eagle clutching arrows and olive branch "U.S./SP RINGFIELD" (note gap in letters) "1864" at rear of plate
Barrel	Underside at rear: Serial number, from "M/1" to "M/3025+"
Furniture	Butt plate: "U.S." on tang Bands (3): "U" near upper edge, right side Rear Sight: "3," "5" only on right rear face of leaf
Stock	"ESA" in oval, on left flat. No date
Principal Dimensions	
Overall Length	52-1/8 inches
Stock Length	48-7/8 inches (band shoulders spaced 9-9/16 inches apart)
Barrel Length	35-9/16 inches; three grooves, 1 turn in 42 inches to the right
Muzzle Diameter	0.775 inch
Ramrod Length	35-15/16 inches (flush with muzzle when stowed) (See **NOTE:** in Ramrod section below)

* The term "blackened" is used to describe the color achieved by case-hardening in oil and water. The oil was floated on water. When the part to be case-hardened was removed from the oven, it was immediately immersed in the oil-water bath. The oil produced the black color and the water hardened the steel.

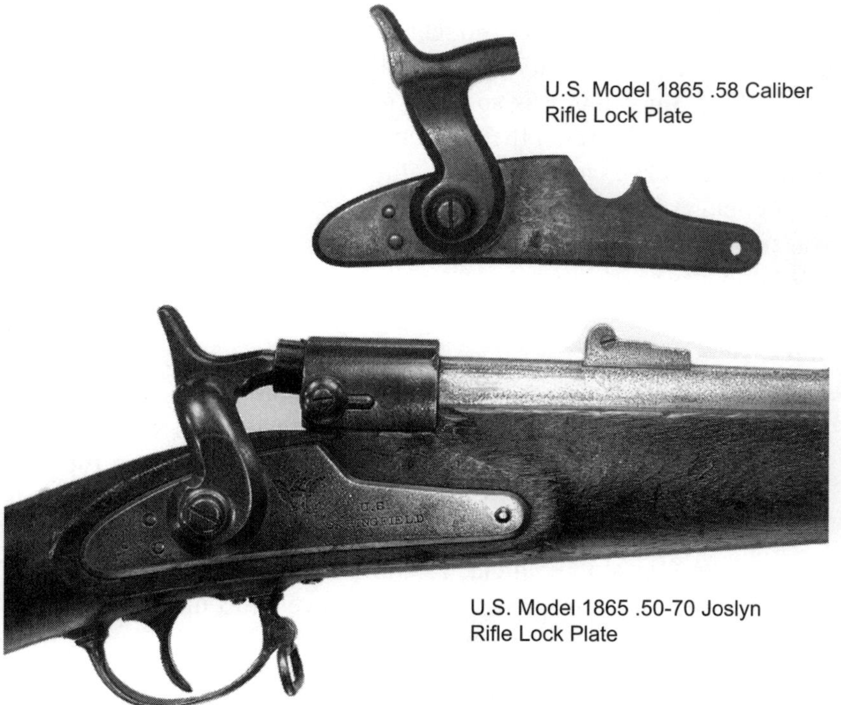

Fig. 10-2. Close-up view of the Joslyn action. Compare the modified lock plate shape to the U.S. Model 1865 lock plate (inset).

Identifying Features

Stock

The stock was entirely different in configuration to other Springfield-made rifles of the period around the breech area (inletting for lock plate, etc.). It even had a slightly different wrist profile from any other Springfield stock. Also of particular notice was the rather severe-looking "drop." However, when compared to the drop of the U.S. Model 1865 stock (1-7/8 inches), it was only 7/8 inch (0.88 inch) more when measured from a line drawn along the top edge of the barrel inletting above the stock heel. So, while there definitely was a difference, the appearance of being excessive is an optical illusion.

It is difficult to believe that anyone taking the time to look at the stock at all could ever have thought it to be any sort of conversion of an existing arm. Stock length, including butt plate and tip, was 48-7/8 inches. Since this was so close to the more usual 48-3/4 inches of the U.S. Model 1868/1870 Springfield-made stocks, the measurement was checked carefully. Also, the use of three barrel bands on a Springfield Armory–made rifle of this overall length (52-1/8 inches) was unique to this rifle. The shoulders for the bands were spaced only 9-9/16 inches apart.

Lock Plate

The lock plate was identical in design, function, and related internal parts, to a normal Springfield-made U.S. Model 1861 or 1863 rifle musket lock, except for one purely cosmetic change. The rounded notch needed for the drum of the musket (and the thumb latch on Allin-system arms) was eliminated by omitting metal in a straight line from the bottom of the old cut, forward to the top of the front curvature. This newly formed edge was then beveled to match the rest of the plate, see Figure 10-3, arrow A.

It should also be noted that these plates were either made new for this arm, or, were modified at some time prior to completion of manufacture, because they have a noticeable flaw in their stamping (SP RINGFIELD) which was not observed on any other arm of the period (arrow B). The date on the plate will always be "1864." Compare this to the U.S. Model 1865 .58 Caliber Springfield Rifle, where many of the lock plates were dated "1865."

Fig. 10-3. Close-up view of the upper edge of the reworked lock plate (A) showing the unusual "SP RINGFIELD" (B) stamping.

Hammer

Fig. 10-4. The U.S. Model 1865 .50-70 Joslyn rifle hammer had a unique shape to its neck; this was an indication of the extreme offset required to strike the centrally mounted firing pin.

The Joslyn hammer was similar to the standard U.S. Model 1863 rifle musket hammer, but was manufactured new, or else extensively reforged to shape, see Figure 10-4. It had a unique "inverted comma" shape and was flattened on the exposed side. It also had an exaggerated (1.115 inches) offset to the left, see Figure 10-5. This offset was needed because of the in-line firing pin, which was located on the centerline of the bore.

Action

The mechanism was quite simple, and very strong, though it lacked a fully developed ejection system. A short, but massive, cylindrical breechblock, or "cap" (Figure 10-5) surrounded the upper portion of the receiver, the front of which pro-

Fig. 10-5. Close-up of Joslyn breechblock from above. Note extreme hammer offset (A), firing pin screw (B), serial number (C), and gas-escape hole (D).

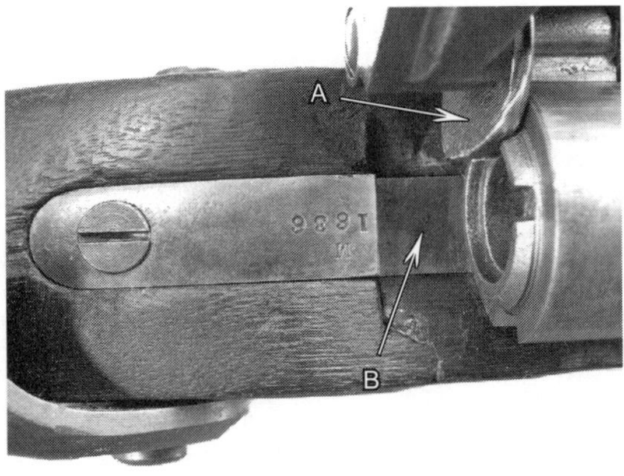

Fig. 10-6. This view shows the extractor plate lip (A) and the locking groove (B) in the tang.

vided a semicircumferential locking surface. In Figure 10-6, it can be seen that a lip on the rear of the cap dropped into a locking groove in the top of the tang.

After bringing the hammer to half-cock, the shooter pulled out the small knurled knob on the right side of the cap about 1/16 inch (0.06 inch). This allowed the cap to pivot on its mounting screw, parallel to the axis of the bore, on the very short (0.552-inch) "receiver," see Figure 10-7.

Fig. 10-7. With the block open, the "swiping" action of the extractor plate (arrow A) can be observed. The angled cut (arrow B) on receiver ring is for the latching mechanism.

A bright-finished semicircular beveled plate, see Figure 10-8 (arrow), was screwed to the front face of the cap in such a manner as to slide beneath the rim of a cartridge being seated.

When the block was swung open, Figure 10-9, this plate (arrow) cammed the case rim back about 7/32 inch (0.22 inch), thus freeing it from the breech. But, then the shooter had to use his fingers to remove it from the rifle, or else shake the rifle to allow it to drop out.

Fig. 10-8. The extractor plate (arrow) which is beveled on both faces, can be seen here, in its notch. Note also the exaggerated "drop" to the stock (the barrel had to be raised in the wood to allow the hammer to strike the firing pin).

When the cap was closed, another camming surface (which formed part of the face abutting the barrel) provided sufficient force to seat a cartridge, though there were some problems reported in chambering cartridges with excessively thick rims.

The spring-loaded firing pin was enclosed in a round tube which projected at a very slight upward angle from the rear of the cap, refer to Figure 10-2. The firing pin retaining screw and a 0.230-inch-diameter gas-escape hole were located on top of this tube, refer to Figure 10-5 (B). The cap was oil-blackened, and was marked on its rear face: "B.F. JOSLYN'S PATENT/OCT 8 1861 JUNE 24 1862," see Figure 10-10.

Serial Numbering

Fig. 10-9. The breechblock face of this (converted centerfire) specimen shows the patched rimfire firing-pin hole (arrow).

The U.S. Model 1865 Joslyn was serially numbered, in a two-line format, consisting of a large "M" above a number (to circa serial number 3,025) in four places; only two such locations are immediately visible, those being (1) the top of the cap, refer to Figure 10-5 (C), and (2) the top of the tang, refer to Figure 10-6. The other locations are (3) the underside of the barrel, just ahead of the receiver joint, and (4) on the concealed face of the extractor cam-plate. On original (.50 Joslyn rimfire) rifles, these numbers should all match. On the 1,600 converted (.50-70 centerfire) rifles, mixed numbers will be the norm, except that the tang should still match the barrel, as these parts did not require disassembly as part of the conversion process.

Fig. 10-10. The patent dates for the U.S. Model 1865 .50-70 Joslyn action were stamped on the rear of the breechblock or cap.

Barrel

The barrel length was 35-9/16 inches, as measured through the bore. It was chambered for either the .50-60 Joslyn or the .50-70 centerfire cartridge in the 1871 alteration. A small notch will be noted at top center of the breech face. This was ground to accommodate the rimfire firing pin and it was not filled in or otherwise eliminated when the rifles were altered to centerfire. Rifling twist was the same for either variation—three broad lands and grooves, having one turn to the right in 42 inches.

The Model 1865 Joslyn barrel had a slightly different contour than the barrel used on the U.S. Model 1863 Springfield rifle musket (or the U.S. Model 1865 .58 Rifle). The breech area was smaller in diameter and this difference was another reason why specially made stocks and barrel bands were needed.

The muzzle diameter (0.775 inch), and placement of front sight stud (1.255 inches behind the muzzle) were the same as that on the U.S. Model 1863 rifle. A standard Model 1855 bayonet fits perfectly, but the correct bayonet had, according to a noted authority, a 21-inch blade to compensate for the shorter (than musket) overall length.

Rear Sight

Fig. 10-11. The U.S. Model 1865 .50-70 Joslyn rifle rear sight was almost identical to the U.S. Model 1863 rifle musket rear sight.

The rear sight was located 2-1/8 inches forward of the breech cap, and was almost identical to the U.S. Model 1863 rifle musket pattern, see Figure 10-11. It was attached to the barrel in dovetail at the rear and secured with a single spanner-head screw that threaded into the barrel. The leaf, with one aperture for 300 yards and notch at top for 500 yards, folded forward. The pivot screw that held the leaf in place entered from the right side rather than from the left as on all other Springfield-made arms of the period.

Furniture

The Model 1865 Joslyn made use of the U.S. Model 1863 rifle musket butt plate, band springs, nose cap, and trigger guard assembly with riveted sling swivel. However, three barrel bands were used to fasten the barrel to the stock, making this the only Springfield arm of this (shorter) length to be so equipped. The forward sling swivel was fastened to the middle band. The bands may appear to be the same as those used on the U.S. Model 1864 rifle musket, but are in fact just slightly smaller to fit the reduced barrel diameter. They do not fit the Model 1863 rifle musket nor any other Springfield-made rifle.

Ramrod

Something of a controversy exists regarding the U.S. Model 1865 Joslyn ramrod. One scholar wrote, in 1972, that he had never seen a Joslyn with its (proper) ramrod. Another, who owned near-mint specimens of both types, naturally disagreed.

The .58- and .50-Caliber Rifles

The author's specimen (a centerfire alteration) has a ramrod, 36 inches long with a constant diameter of 0.182 inch and a Springfield-like 7-ribbed head. But, because of a clearly different patina, this rod cannot be original to the rifle, and is thus suspect. The stock contains a threaded plate to accept a screw-in attachment (as in the U.S. Model 1866 Rifle). Other owners consulted have reported similar inconsistencies, involving length, diameter, head shape, and mode of attachment.

NOTE: The most accurate information regarding the ramrod to date seems to be:

Head shape: same as the U.S. Model 1866 Springfield Rifle
Length: rimfire version, 35-15/16 inches; centerfire version, 35-11/16 inches
Diameter: rimfire version, 0.193 inch to 0.189 inch; centerfire version, 0.188 inch to 0.18 inch
Small End: threaded #12-26

The differences in length and head diameter between the rimfire and centerfire measurements may simply be the result of manufacturing tolerances as opposed to design differences. The author requests any information the reader might be able to provide from an example in his or her collection.

Chapter 11: The U.S. Model 1868 Springfield-Remington (.50-70) Transformed Rifle

Historical Background

Leonard Geiger patented the original version of the "Remington System" in 1863. These first actions, known as the "split-breech model," had a distinctly different profile as seen from the side than later rolling block actions. They were used on rimfire carbines ordered by the Government at the very end of the war.This basic action was subsequently modified, simplified, and greatly strengthened by the work of Joseph Rider, a Remington employee, who patented improvements in 1866, 1867, and 1871.

The so-called "No. 1" receiver was the largest rolling block action made, with well over one million fabricated at the Remington factory in Ilion, New York, and in various countries abroad under license. At one time it was the official military rifle of at least seventeen nations, ranging from Northern Europe to East Asia to South America. These foreign rifles and carbines can still be found in huge numbers today, and at reasonable prices. Their numerous variations provide an almost endless choice for the military arms collector and/or shooter. But, by far, the most highly prized rolling block models of all are those U.S. military versions, produced at the Springfield Armory.

Quantity Produced

On November 7, 1867, Brigadier General A. B. Dyer, the Chief of Ordnance, ordered 500 No. 1 rolling block actions from Remington at $7.00 each. A few

Fig. 11-1. The U.S. Model 1868 Springfield-Remington .50-70 (Transformed) Rifle in its original, "long" configuration with the 39-3/8-inch barrel. Photograph courtesy of Springfield Armory NHS, SPAR #5901.

days later, he ordered the Springfield Armory to convert 504 U.S. Model 1864 rifle muskets to breechloading rifles using the Remington actions. Remington reworked the rifle musket barrels and attached them to the actions. The work was completed in the spring of 1868 and they were shipped to the Springfield Armory, where the rest of the conversion process was accomplished. They were referred to originally as the "Remington Breechloading Muskets," but were officially called "Springfield-Remington Transformed Rifle Muskets (Long) " due to their 39-3/8-inch-long barrels, see Figure 11-1. Later, almost all had their barrels shortened to 36 inches at the Springfield Armory in August 1869, as outlined below, and were renamed the "Springfield-Remington Transformed Rifle Muskets (Short)," see Figure 11-2. The original models as converted (long version) were not serially numbered, but a serial number from 1 to 504 was added to the top of the barrel, at the time they were shortened.

Fig. 11-2. The U.S. Springfield-Remington .50-70 (Transformed) Rifle in its "short" configuration with the 36-inch barrel. Note that only two barrel bands were used. Ed Hull collection.

Adapted From

These conversions were created from the parent U.S. Model 1864 rifle musket by literally cutting the rifle musket stock in two, and splicing in the No. 1 rolling block action. The barrels were remachined and fitted to the action at Remington.

Table 11 The U.S. Model 1868 Springfield-Remington (.50-70) Transformed Rifle	
Finishes	
Receiver	Color Case-Hardened
Breechblock or Bolt	National Armory Bright

Hammer	National Armory Bright
Barrel	National Armory Bright
Furniture	National Armory Bright Rear Sight: Browned (blued)
Stock	Oil-finished American Black Walnut
Markings	
Receiver	Tang: Remington patent markings. This model did not carry the "U.S./ SPRINGFIELD" marking on the right side of the receiver. As altered, Serial #s 1-500 (approx.) at top
Breechblock or Bolt	Thumbpiece: knurled edge to edge
Hammer	Hammer: knurled edge to edge
Barrel	Witness marks at top, rear center As converted: none
Furniture	Butt Plate: "U.S." on tang Bands (3 as converted, 2 as altered): "U" near upper edge, right side Rear Sight: "3," "5" only on right rear face of leaf
Stock	"ESA" and "SWP" in ovals on left flat. No date. "SWP" is not seen on all specimens
Principal Dimensions	
Overall Length	55-1/4 inches as originally converted (very scarce) 51-7/8 inches as altered
Stock Length	Buttstock: 13-3/8 inches (including the butt plate) Forend (as converted, including nose cap): 34-11/16 inches Forend (as altered): 31-5/16 inches
Barrel Length	39-3/8 inches as converted 36 inches as altered
Muzzle Diameter	0.775 inch
Ramrod Length	38-1/8 inches (as converted) 34-1/2 inches (as altered). Flush with muzzle when stowed

Identifying Features

Stock

As noted above, the musket stocks were literally cut into two pieces. The butt portion was inletted for the upper and lower tangs, and profiled to mate with the rear of the new receiver. When finished, it was 13-3/8 inches long, including the butt plate. Two separate cartouches were stamped on the left wrist, "ESA" and "SWP." The "SWP" mark does not occur on all specimens. As these rifles were used heavily, the cartouches may be faint or worn away on surviving specimens, see Figure 11-3.

Fig. 11-3. The U.S. Springfield-Remington Transformed Rifle (short) shows cartouches by both E.S. Allin and S.W. Porter, reflecting their different courses and periods of work at the Springfield Armory. Photograph courtesy of Springfield Armory NHS, SPAR #5901.

The forend was machined into a tenon that fitted into the front of the new action. The exposed length, after remachining, was 34-11/16 inches. The three U.S. Model 1864 barrel bands and band springs were retained when first converted. However, when the barrels were shortened to 36 inches, the forend had to be shortened as well and only two of the barrel bands were remounted. The unused spring slot was patched with a strip of walnut, and inletting was added for a Model 1868 ramrod stop. These reused forends were thicker than those made for the later U.S. Model 1870 Navy Rifle (Chapter 13) and U.S. Model 1871 Army Rifle (Chapter 15).

Barrel

The existing 40-inch rifle musket barrel was turned down and threaded to suit the new Remington actions. This reduced the actual barrel length to 39-3/8 inches long as measured through the bore. The barrel was then bored out and fitted with a brazed-in liner to reduce the caliber from .58 to .50 inch, exactly as had been done on the U.S. Model 1866 Rifle. Doing so removed the internal threads formerly filled by the tang screw. The new liner was rifled with three broad grooves, having a twist of one turn in 42 inches, and was chambered for the .50-70 cartridge.

The one-piece combination front sight/bayonet lug base already on the rifle musket barrel was retained on the Springfield-Remington conversion, see Figure 11-4. This, plus the retention of the original muzzle diameter meant that the converted rifle was able to accept the standard Model 1855 angular bayonet.

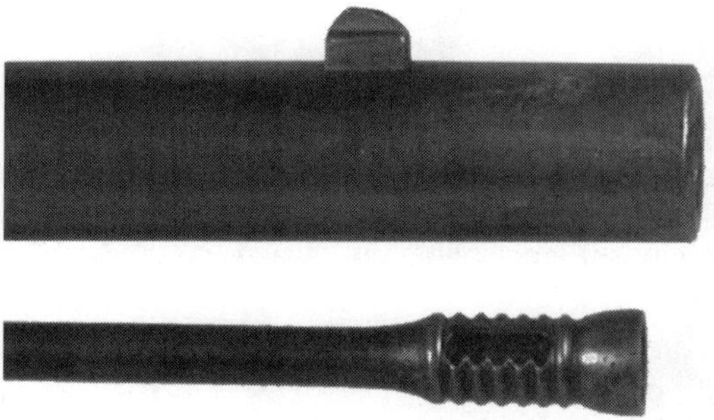

Fig.11-4. The U.S. Model 1863–type one-piece front sight and base was retained on the Springfield-Remington Transformed Rifles. Springfield Armory NHS collection, SPAR #5901.

In August 1869, nearly all of these arms had their barrels reduced to an even 36 inches in length, making an original uncut specimen extremely rare. The shortening process involved cutting off the muzzle end of the barrel and rechambering and rethreading it to fit the receiver. The new name for the converted rifle was changed to "Springfield-Remington Transformed Rifle Musket (Short)."

Action

The familiar Remington cast and machined receiver was used, see Figures 11-5 and 11-6. A new hammer was produced that had the same the same profile as the 1867 and 1870 Navy models (Chapter 13) but was noticeably smaller than that of the 1871 Army model (Chapter 15). The hammer spur jutted to the rear at a 55-degree angle whereas the later rolling block hammer spur was set at a 40-degree

Fig. 11-5. The Remington receivers used for the U.S. Model 1868 Springfield-Remington Transformed Rifles carried the patent markings on the tang. The hammer was similar in profile to that used on the U.S. Model 1867 and 1870 Remington naval rifles. Springfield Armory NHS collection, SPAR #5901.

Fig. 11-6. Left-side view of the U.S. Model 1868 Springfield-Remington Transformed Rifle hammer showing the similarity to that used on the U.S. Model 1867 Cadet and 1870 Remington naval rifles. Springfield Armory NHS collection, SPAR #5901.

angle. The knurled thumbpiece on the breechblock projected upward. This early design utilized the sliding extractor that was fitted into the left wall of the receiver. Unlike most later Springfield rolling blocks, this arm did not bear any Springfield markings on the receiver. The upper tang carried the Remington patent markings as follows:

REMINGTON'S ILION, N.Y. U.S.A.
PAT. MAY 3D NOV. 15TH 1864, APRIL 17TH 1866

NOTE: Many earlier, purely Remington-made, rolling block actions have the final patent date as "April 17th 1866," which is the date of the last-known Geiger patent. However, some later Remington guns, notably the New York State contract arms of 1871, include the date "April 17th 1868." Further, some other "pre-Springfield" examples are known where the final digit appears to be overstruck with an "8." The reason for this is unknown. Was there a second patent? Ed Hull, who has made a study of American arms during the period of transition from percussion to self-contained metallic cartridges, has suggested that perhaps Springfield Armory workmen unwittingly copied, and perpetuated, an erroneous marking. In any event, the Springfield-Remington tangs are marked with a date which is exactly two years, to the day, later than the date of the Geiger patent reissue. Collector input is solicited.

Rear Sight

The original U.S. Model 1864 Springfield rifle musket rear sight was retained without change, refer to Figure 11-7.

Furniture

Virtually all of the furniture but the lock and trigger assemblies were reused to build the "transformed" arms with the 39-3/8-inch barrel. This included the three barrel bands and band springs from the parent U.S. Model 1864 rifle musket, see Figure 11-8 and refer to Figure 11-1. When the barrels were shortened to 36 inches the following year, only two bands of the original three were remounted. Compare Figures 11-1 and 11-2. The upper band carried the forward

Fig. 11-7. The U.S. Model 1864 Springfield rear sight was mounted on the Springfield-Remington Transformed Rifles. Springfield Armory NHS collection, SPAR #5901.

sling swivel. The rear sling swivel, which was mounted on the trigger guard plate, was retained. As noted above, the nose cap, and the butt plate marked "U.S.," were reused from the surplus Model 1864 rifle musket inventory.

Fig. 11-8. The barrel, barrel bands, band springs, and nose cap from the U.S. Model 1864 rifle musket were all reused on the Springfield-Remington Transformed Rifle. Unlike the .58-caliber Remington conversions which retained the "tulip" rod, this arm used a Model 1866–type rod. Springfield Armory NHS collection, SPAR #5901.

NOTE: While the U.S. Army did not order any additional U.S. Model 1868 Springfield-Remington Transformed Rifle Muskets beyond the original 504, the state of South Carolina ordered 5,000 similar rifles chambered for a .58-caliber centerfire cartridge. These rifles were designated the "Remington Transformed Rifle Muskets." Another 1,000 were made for the state government of Texas. An unknown number of additional rifles of both the long- and short-barreled pattern were manufactured by Remington and sold to other customers. These were designated as "Springfield Model—Caliber .58—Angular Bayonet" for the 39-inch-barrel rifle and "Springfield Model—Cal. .58 (Short)—Angular Bayonet" for the 36-inch-barrel length in the Remington catalogs of the period. Both Springfield rifle musket and very early Enfield rifle musket parts were used in these conversions. Most were chambered for the .58-caliber centerfire cartridge but an unknown number were chambered also for the .58-caliber rimfire cartridge.

Ramrod

At time of original conversion, it is thought that the existing tulip-headed rifle musket ramrod was simply reused and its lower end was threaded into a stud inserted in front of the receiver to prevent it from flying forward under recoil. However, the example of the U.S. Springfield-Remington .50-70 Transformed Rifle (Long) in the Springfield Armory Museum (SPAR #5901) is equipped with a Model 1864- or Model 1866-type ramrod, see Figure 11-8. When the arms were shortened in 1869, a single-shoulder Model 1868-type ramrod was added, with its attendant ramrod stop, which was secured by the upper band-spring pin.

Chapter 12: The U.S. Model 1867 .50-45 Navy Cadet Rifle (Remington Patents)

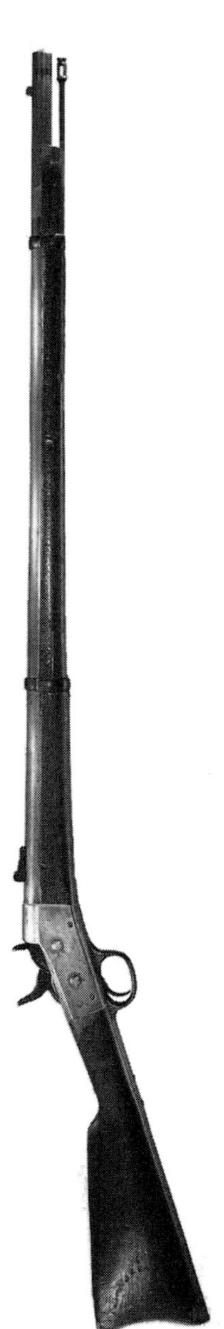

Historical Background

The U.S. Model 1867 .50-45 Cadet Rifle, ordered by the U.S. Navy, was the only rifle ever designed specifically for the midshipmen at the U.S. Naval Academy at Annapolis, see Figure 12-1. Prior to adoption of this model, the USNA cadets had been equipped with a hodgepodge of altered full-size rifles, and "army surplus" percussion cadet arms. After its adoption, the Model 1867 Navy cadet rolling block rifle enjoyed the longest tenure (some 22 years), in unaltered form, of *any* Springfield-produced arm of the period! They were not replaced at the academy until around 1890, at which time they were exchanged for the U.S. Hotchkiss Navy Model Rifles. The rather worn condition of most surviving examples reflects this long usage.

Quantity Produced

Just 498 of these rare Model 1867 Navy Cadet Rifles were made during mid-1868.

Adapted From

The Model 1867 Navy Cadet Rifle represented the second effort by the Springfield Armory to assemble an all-new arm using the Remington No. 1 action after their recent involvement with the conversion of 504 Model 1864 rifle muskets to the Remington system (see Chapter 11). It used the same action as the Model 1867 Navy Carbine, which was fabricated completely by Remington.

Fig. 12-1. U.S. Model 1867 .50-45 Navy Cadet Rifle. Note the distinctive scaled-down buttstock.

In fact, it was once thought that these cadet rifles may have been made using actions obtained by disassembling some of the Model 1867 carbines, but this is now known to be incorrect, as the carbines remained in Navy hands until well after the Cadet models had been produced. Further, it has been observed that the inspector's marks were applied differently. For instance, the Navy anchor was stamped on the left side of the Cadet receiver, but on the right side of the Navy Carbine.

In overall profile, this rifle was quite similar to the later Model 1870 Navy Rifles, see Chapter 13. However, this was one of only two arms from this time period (the other being the Allin-system U.S. Model 1866 Cadet rifle) wherein an entire rifle was meticulously scaled down, part by part, to fit the smaller stature of its intended users. The length of pull, at only 12-5/16 inches from the center of butt plate to center of trigger, was 1-5/16 inches shorter than a standard-sized rolling block rifle. The difference is quite noticeable to an adult shooter, when shouldering the piece.

Table 12A The U.S. Model 1867 .50-45 Navy Cadet Rifle (Remington Patents)	
Finishes	
Receiver	Color Case-Hardened
Breechblock or Bolt	National Armory Bright
Hammer	National Armory Bright
Barrel	National Armory Bright
Furniture	National Armory Bright Rear Sight: Blackened
Stock	Oil-finished American Black Walnut
Markings	
Receiver	Upper tang: Remington patent markings. Right side center bottom: "P/F.C.W." Left side center bottom: Anchor (The receiver did not carry the "U.S./SPRING-FIELD" marking)

Breechblock or Bolt	Thumbpiece: knurled edge to edge
Hammer	Hammer: knurled edge to edge
Barrel	Witness marks at top, rear center
Furniture	Butt Plate: "U S" on tang (without periods) Bands (2): "U" near upper edge, right side Rear Sight: "3," "5" only on right rear face of leaf
Stock	"ESA" in oval on left flat. No date
Principal Dimensions	
Overall Length	47-5/16 inches
Stock Length	Buttstock: 12-1/4 inches Forend (including nose cap): 27-5/8 inches
Barrel Length	32-9/16 inches; 3 grooves, 1 turn in 42 inches to the right
Muzzle Diameter	0.735 inch (used the Model 1866 Cadet Bayonet)
Ramrod Length	31-9/16 inches (flush with muzzle when stowed)

Identifying Features

Stock

As stated above, this Cadet model was scaled down from the full-size rifle. The butt portion of the stock was only 12-1/4 inches long, measured from the heel to the lower corner of receiver, compared to 13-1/2 inches for the full-size rifle. The circumference of the wrist, at 4-3/4 inches, was 1/2 inch less than normal, see Figure 12-2.

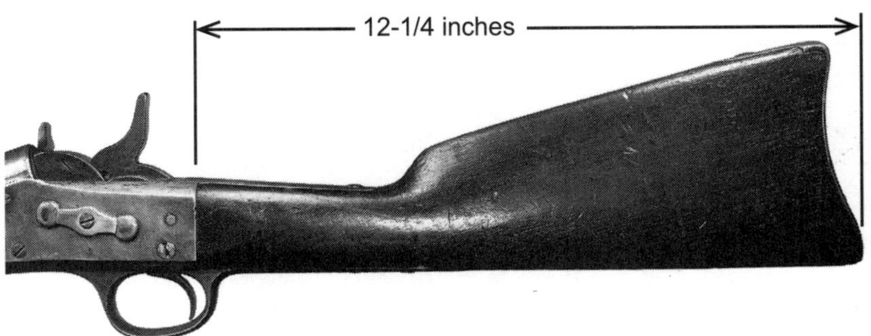

Fig. 12-2. This view of the buttstock shows the U.S. Model 1867 Navy Cadets' slim wrist and reduced length.

The exposed portion of the forend was 27-5/8 inches long, including tip, compared to 31-5/16 inches for the full-size rifle, and was noticeably thinner by 1/8 inch (0.13 inch) than all other rolling block arms. The band shoulders were just 16-1/8 inches apart.

Future cadet rifles manufactured at the Springfield Armory, beginning with the Allin-system U.S. Model 1869 Cadet rifle, would do away with this complication and simply shorten the barrel (between bands) and slim the butt slightly, keeping length of pull standard, thus staying more compatible with the infantry rifle.

Barrel

The barrel length was 32-9/16 inches and it was chambered for the short .50-45 Carbine round, described below. In most other respects, the barrel was similar to that used on the Model 1870 Navy rifle which followed. The front sight was the typical Springfield one-piece combination front sight and bayonet lug. The barrel was rifled with three broad grooves, having a twist of one turn in 42 inches on the specimen examined. A number such as "100" was stamped on the underside of the barrel, just ahead of the receiver.

Fig. 12-3. Two inside-primed .50-45 cartridges. Left, Martin "bar" primed (1.275-inch case); right, Benet primed (1.332-inch case). Both cartridges were used in this Cadet rifle.

Like the 1866 Allin conversion, the barrel was remade from a Model 1863 rifle musket barrel reduced in length, and also reduced in caliber from .58 to .50 inch with a brazed-in liner. The exterior of the barrel was tapered to a muzzle diameter of 0.735 inch to accept the special Model 1866 Cadet socket bayonet.

Cartridge

The U.S. Model 1867 .50-45 Navy Cadet Rifle fired a special, short case .50-45 centerfire cartridge see Figure 12-3. The copper-case held a powder charge of 45 grains of FFg black powder. The rather pointed lead bullet weighed 430 grains. The cartridge was manufactured at Frankford Arsenal using both major case/primer types—the Martin bar anvil, and Benet cup primer. Neither of these early .50-caliber military cartridges had a headstamp.

A typical specimen of the Martin bar anvil type, from the author's collection, measures as follows. Rim: 0.661 inch by .0580 inch thick; case: 0.563 inch at head, 0.525 inch at mouth, case 1.275 inches long. Overall length was 1.750 inches. The Benet cup-primed round was a bit longer (case: 1.332 inches, and overall: 1.768 inches) but both rounds chambered easily in the Model 1867 Navy Cadet rifle. The chamber was purposefully cut only to this size, and the rifle will not accept a full-length .50-70 cartridge.

Rear Sight

The rear sight base used on the Cadet rifle was taken from surplus U.S. Model 1863 rifle muskets, see Figure 12-4. The rear sight was mounted on the barrel 3/4 inch ahead of the receiver, in a dovetail and secured with a spanner-head screw. The sight had one leaf, hinged at the rear to fall forward. The leaf had a hole, marked "3" for 300 yards, and a notch at the top, marked "5" for 500 yards.

Fig. 12-4. Surplus U.S. Model 1863 rifle musket rear sights were installed on the U.S. Model 1867 Navy Cadet.

Action

The hammer, see Figure 12-5, was noticeably smaller than that used on the more common Remington Model 1871 Army model, described in Chapter 15. The spur leaned back 55 degrees from the vertical. The knurled thumb-

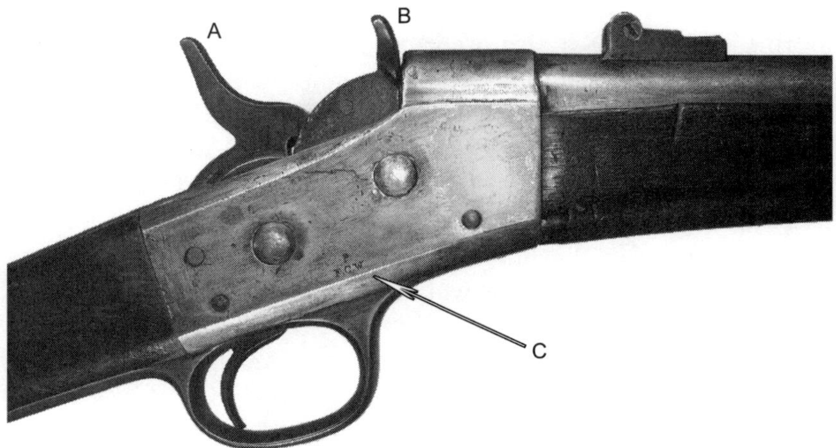

Fig. 12-5. The right side of the U.S. Model 1867 Navy Cadet receiver. Note the shape of the hammer (A), orientation of the breechblock thumbpiece (B), location of the inspector's initials and proof mark (C) just above the front of the trigger guard bow.

piece on the breechblock projected upward, and its right edge had a different profile than subsequent models. This early version of the rolling block design utilized a spur extractor set into a slot in the breechblock hinge knuckle, see Figure 12-6, arrow.

Fig. 12-6. This view shows the detail of the spur extractor (arrow), mounted in the breechblock. The spur sometimes tore through the soft rim of the copper-alloy cartridge cases then in use.

Prior to ordering the U.S. Model 1867 Cadet Rifle, the Navy had become very unhappy with their Remington carbines in service, largely due to this extractor. It was intended to catch the bottom of the cartridge rim and lever it out of the chamber. In fact, it often tore through the soft copper rim.

In fairness, this was probably as much an ammunition problem as a "rifle" problem as it also occurred in the early .45-70 Springfield rifles, and it continued to be a problem until the use of copper cases was ended by the Ordnance Department in the mid-1880s. Recognizing the existence of this problem, Remington had wanted the Navy to accept their newly designed sliding extractor for the Model 1867 Navy Cadet Rifle, and even went so far as to provide the hole for its retaining screw. But the Navy rejected the offer! The hole was plugged with a dummy screw which served no function.

At that time, the rifle sling was not in use at Annapolis and so sling swivels were not mounted on the Model 1867 Navy Cadet Rifle.

As noted under Markings, the upper tang was stamped with Remington patent dates. They were as follows: "REMINGTON'S ILION N.Y. U.S.A./PAT. MAY 3^{D} NOV. 15TH 1864 APRIL 17TH 1866," see Figure 12-7.

Fig. 12-7. The Remington patent dates were stamped on the upper tang. Compare these to the patent dates shown in Figure 13-5.

The receiver was marked on the left side (center bottom) with the Navy's acceptance mark, which was an anchor, and on the right side and at the center bottom with the proof mark, "P," above "F.C.W." for Frank C. Warner, the Navy inspector, see Figures 12-8 and 12-9. This was a different format than that used on the previous carbine, where the Navy acceptance anchor appeared on the right side, just below the initials.

Even though this arm was assembled at Springfield, and many of its special parts were made there, it (unlike most other Springfield-Remington models) does not bear the Springfield name. Reportedly, the

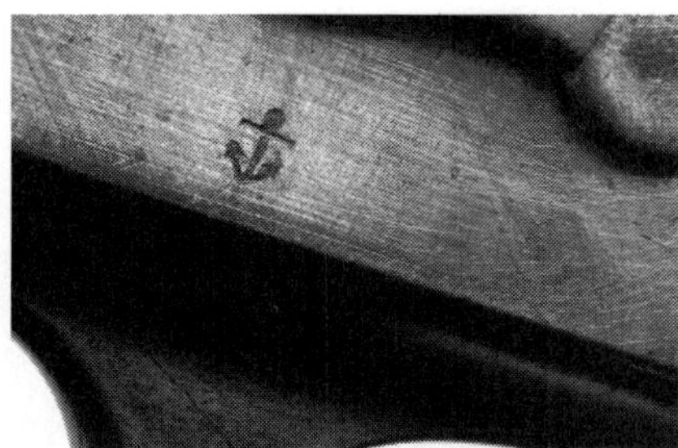

Fig. 12-8. The Navy's acceptance stamp was an anchor stamped on the left side of the receiver, just above the front of the trigger guard.

Fig. 12-9. This view shows Navy inspector Frank C. Warner's initials on right side of receiver, at the bottom edge, just above trigger guard. Above his initials is the "P" proof mark.

sequence of manufacturing and assembly operations went like this: Springfield shipped finished barrels to the Remington factory at Ilion, NY, where they were fitted to the actions. The barreled actions were then returned to Springfield where the rest of the parts involved were manufactured and the final assembly into a finished arm completed. It should be noted that no surplus Civil War rifle musket parts were usable in this rifle, except the barrel and rear sight.

Furniture

In keeping with the significantly shorter length of the rifle, a number of parts were reduced in size, and are thus not interchangeable with other contemporary Springfield arms. A few parts such as the butt plate (see Figure 12-10), barrel bands (see Figure 12-11) and band springs were interchangeable with the Model 1867 Army Cadet Rifle. The table below provides dimensional comparisons of certain Navy Cadet parts to the full-size Model 1870 Navy Rifle.

Fig. 12-10. The butt plate tang was marked "U S" (without periods) the only U.S. government ownership mark on the rifle. Below it is the Naval Academy's engraved alphanumeric rack number.

Fig. 12-11. The U.S. Model 1867 Navy Cadet rifle upper band was so small, relative to the .50-caliber barrel, as to almost give the appearance of a perfectly round ring of metal.

Table 12B
Comparison of Selected Parts
U.S. Model 1867 Navy Cadet Rifle and Model 1870 Navy Rifle
(inches)

Rifle Model	Navy Cadet	Navy Service
Butt plate: width/length	1.50 by 4.25 (1867)	1.69 by 4.5 (1870)
Barrel bands: width, length, thickness		
Lower	1.35 by 1.72 by 0.08	1.79 by 1.42 by 0.09
Upper	1.10 by 1.45 by 0.08	1.78* by 1.22 by 0.09
Band springs: length exposed	1.78	2.01
Nose cap: width, length, thickness	0.92 by 0.80 by 0.74	0.99 by 0.95 by 0.86
* Including swivel mounting		

With the perfect hindsight of over 135 years, the foregoing reduction in part sizes for such a small run of rifles might seem ridiculous. The Navy could not possibly have saved enough on weight to have overcome the nuisance of making and keeping track of the special parts, particularly the band springs! In closing, it should be noted that all of the above parts, even though they were gauged, are subject to slight variations in size from the normal tolerances of grinding and polishing.

Ramrod

The ramrod threaded into a stud at the front of the receiver. It was smaller in diameter than that for the full-sized U.S. Model 1866 Springfield Rifle, tapering from 0.211 inch behind the head to 0.194 inch at the lower end, which was threaded with 26 threads to the inch. The length of the ramrod was 31-9/16 inches. The head had seven rings and a cupped face, and was flush with the muzzle when stowed. This rod is similar to that for the U.S. Model 1866 Springfield Cadet (Chapter 5) except for diameter and length. The Navy rod is exactly one inch shorter, see Figure 12-12.

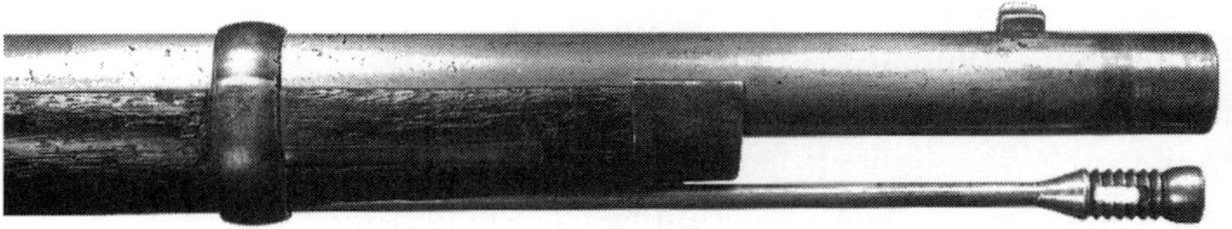

Fig. 12-12. The slender ramrod used in the U.S. Model 1867 Navy Cadet rifle is apparent in this view. The band spring and nose cap are also noticeably smaller than normal.

Chapter 13: The U.S. Model 1870 Navy Rifles, Types I and II (Remington Patents)

Historical Background

Although the rest of the world purchased, and/or fabricated under license, well over a million military rifles using the Remington/Rider/Geiger rolling block patent design, it never truly caught on with the U.S. Army, though they tested more than one variation in several trials. The U.S. Navy did issue them for general use, but not in truly large quantities.

Following the less-than-successful trial of the 5,000 Model 1867 Remington-manufactured carbines, the Navy ordered 10,000 of this particular model in 1870 in .50-70 caliber, but upon their completion, complained of a flaw in the design, and refused to accept them. The problem had to do with the rear sight. It has long been held that it was simply the rearward location just 1/2 inch (0.5 inch) ahead of the receiver that was the problem; the author believes it was at least as much the manner of mounting that caused the trouble.

This was the first use of the U.S. Model 1868 "long-range" sight on any type of rifle than an Allin-system "trapdoor." To provide support to the sight leaf when in the down position, the barrel was ground down, on the top only, in a reverse taper, to provide a "swamped" bed, or seat, for the sight so that the leaf could lie flat on the receiver.

Fig. 13-1. The U.S. Model 1870 Navy Rifle in .50-70 caliber. This is the Type II rifle with the Remington patent breech. The Type I is identical except for location of rear sight.

The Navy was convinced that the barrel was unduly weakened by this uneven removal of metal so near the chamber. Coincidentally, the rearward mounting location also permitted duplicating the sight radius of the Allin arm, thus eliminating any need to revise the gradations. The Springfield Armory later solved the same issue with this sight by inserting a special screw at the top of the leaf. The screw served as a pin which rested against the barrel and supported the leaf, preventing it from being bent or cracked.

In any event, the Navy sold the rejected rifles to the firm of Poultney & Trimble. P&T purchased 2,178 rifles directly from the Navy and an additional 7,822 more directly from Springfield Armory. Congress, suspecting that "someone" had profited from this transaction, conducted an investigation in 1871 and discovered that someone had, the U.S. Navy. They realized enough from the sale of the rifles to have Springfield Armory manufacture an additional 12,000! P&T resold the rejected rifles to Remington who in turn sold them to France.

The 12,000 new rifles corrected the rear sight problem; it was relocated 3-1/4 inches ahead of the receiver, see Figure 13-1. There are those who claim that the whole matter was a carefully orchestrated scam by the Navy to gain an additional 2,000 rifles and that there was nothing really wrong with the first production run.

Quantity Produced

A total of 22,013 rifles were made: 10,000 "Type I" rifles with the rear sight 1/2 inch ahead of the receiver, and 12,000-plus "Type II" rifles with the rear sight 3-1/4 inches ahead of the receiver. The later Type II version, with forward-placed rear sight, is encountered much more frequently by collectors today. A total of 1,302 Type I rifles had entered service before the model was rejected. A full military specimen of the "Type I" arm, with U.S. Navy proof marks and with the bayonet lug intact, is considered a very desirable find.

Adapted From

A typical Remington "No. 1" black powder action was made under license at Springfield, as the Remington plant had prior manufacturing commitments. Although it was reported to have been somewhat redesigned at Springfield, except for a change in extractor design it was very similar to the receivers used on the previous 1867 Navy Cadet rifles.

Table 13 **The U.S. Model 1870 Navy Rifles,** **Types I and II (Remington Patents)**	
Finishes	
Receiver	Color Case-Hardened (including trigger guard)
Breechblock or Bolt	National Armory Bright
Hammer	National Armory Bright
Barrel	Type I, National Armory Bright. Type II, Blued
Furniture	Butt Plate and Nose Cap: National Armory Bright Rear Sight: Browned (blued) Barrel Bands: some will be found blued
Stock	Oil-finished American Black Walnut
Markings	
Receiver	Upper tang: Remington patent markings. Right side: Eagle clutching arrows and olive branch/"USN/SPRINGFIELD/1870"
Breechblock or Bolt	Thumbpiece: knurled edge to edge
Hammer	Hammer: knurled edge to edge
Barrel	"P/H.B.R." at left rear top Anchor and witness marks at center rear top
Furniture	Butt Plate: "U.S." on tang Bands (2): "U" near upper edge, right side Rear Sight: "2," "3," "5," "7," "9" only on rear face of leaf
Stock	"ESA" in oval on left flat. No date
Principal Dimensions	
Overall Length	48-9/16 inches

Stock Length	Buttstock: 13-1/2 inches Forend (including nose cap): 26-5/16 inches Band shoulders: 14-5/8 inches apart
Barrel Length	32-5/8 inches; 3 grooves, 1 turn in 42 inches to the right
Muzzle Diameter	0.775 inch
Ramrod Length	29-1/2 inches (1-11/16 inches short of muzzle when stowed)

Identifying Features

Stock

The stock was the typical Remington rolling block-style two-piece buttstock and forend made of American black walnut. The buttstock was fitted carefully to the upper and lower tangs and shaped at the butt for the regulation Springfield rifle butt plate. The buttstock was 13-1/2 inches long as measured along the bottom, from the toe to the lower rear corner where it joined the receiver. The stock bore a small oval "ESA" cartouche on the left side of the wrist, the inspection cartouche of Master Armorer Erskine S. Allin.

The forend was 26-5/16 inches long, including the nose cap which was 15/16 inch (0.94 inch) long, and was inletted for two bands. The shoulders for the bands were 14-5/8 inches apart. They were held in place against recoil by a band spring. The rear portion of the forend terminated in a square tenon which fitted into a socket in the front of the receiver. The forend was also inletted for a barrel lug or block which was screwed to the underside of the barrel. This lug prevented the forend from sliding forward under recoil.

The forend also had a ramrod stop for the single-shoulder Model 1868 ramrod, which was secured by the upper band spring. The forend terminated farther back from the muzzle than other Springfield rifles of the period because of the space required for the bayonet lug.

Action

The receiver was marked on right side with a large spread-winged eagle clutching arrows, above "USN" above "SPRINGFIELD" above "1870," see Figure 13-2. The hammer was nearly identical to that used on the U.S. Model 1867 Navy Cadet and was noticeably smaller than the hammer used on the U.S. Model 1871 Army rifle, see Figure 13-3.

Fig. 13-2. The U.S. Model 1870 Navy Rifle is marked on the right side of receiver only. There are no markings on the receiver's left side.

The spur was bent to the rear. The knurled thumbpiece on the breechblock projected upward.

Instead of the extractor lug used on the Model 1867 Navy Cadet, both the Model 1870 Type I and Type II Navy rifles used the sliding extractor which the Naval authorities had earlier rejected, see Figure 13-4. The new extractor was considered a definite improvement.

Fig. 13-3. The U.S. Model 1870 Navy rifle action (above) compared to the U.S. Model 1871 Army rifle action (below). Note the differences in the hammers and thumbpieces.

Fig. 13-4. This view shows the sliding extractor (arrow) incorporated into this model. It was a significant improvement over the earlier lug because it worked in a straight line and contacted nearly twice as much of the case rim.

The lower sling swivel was mounted directly in the front leg of the massive one-piece trigger guard, which was an integral part of the lower tang.

The upper tang, see Figure 13-5, was marked with the Remington patent dates as follows:

REMINGTONS PATENT
PAT. MAY 3$^{\underline{D}}$ NOV. 15$^{\underline{TH}}$ 1864 APRIL 17$^{\underline{TH}}$ 1866

Fig. 13-5. Remington patent dates on upper tang. Note change from the format used on the preceding Cadet model in Fig. 12-7. Last digit is a "6" and not "8."

Barrel

The barrel was 32-5/8 inches long and chambered for the .50-70 caliber rifle cartridge. The bore was rifled with three broad lands and grooves having a twist of one turn in 42 inches. A typical one-piece Springfield Armory front sight was mounted 1-1/4 inches behind the muzzle, but in these Navy rifles, it was not intended to do double-duty as a lug for the angular bayonet. Instead, the Model 1870 Navy rifles were fitted for a sword bayonet. To mount the bayonet, a separate, two-stepped block of metal, 0.542 inch by 1.50 inches, with "T"-shaped lug was mounted on the underside of the barrel, 1.95 inches behind the muzzle, see Figure 13-6.

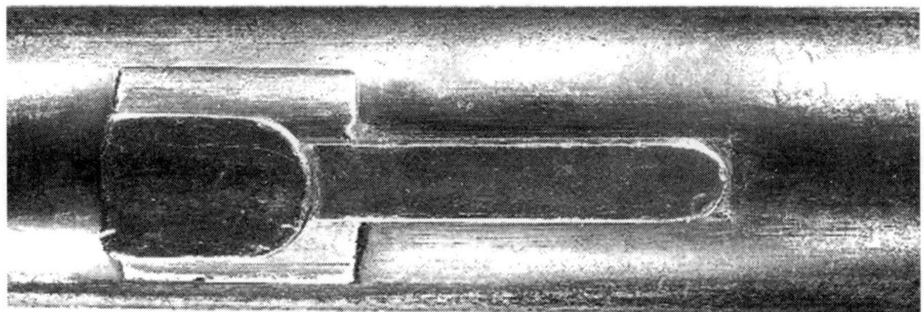

Fig. 13-6. The bayonet lug for the sword bayonet was mounted under the barrel. Some altered U.S. Model 1870 Navy rifles lack this feature which was removed by some commercial surplus dealers.

Interestingly enough, the Model 1870 Navy Rifle will also be found without the bayonet lug. This has been determined to be the work of post-service commercial dealers to allow use of the standard U.S. Model 1855 socket bayonet. Absence of the lug does lessen the desirability, and hence value, of the arm.

The barrel was marked, just ahead of the receiver, on its left side, with the letter "P" above "H.B.R." This stamping on the author's Type I was quite thick, coarse, and deep, see Figure 13-7. The same markings on the Type II were much finer, almost spidery, and lightly struck, see Figure 13-8. An anchor was stamped on the barrel at top center, just in front of the receiver, see Figure 13-9, along with a pair of witness marks to align the barrel in the receiver. On the author's

Type I, the top of the anchor points toward the muzzle; on his Type II, the anchor points toward the breech. It is not known if this is an actual model difference or simply a stamping variation. The author would be grateful for any clarifying information.

Rear Sight

The Model 1868 long-range sight designed at the Springfield Armory was used on both types of the Model 1870 Navy rifle. The sight leaf overhung the sight base by nearly 5/8 inch (0.63 inch). On the first 10,000 Type I rifles, the rear sight was mounted 1/2 inch (0.5 inch) ahead of the receiver. To support the leaf when it was down and to prevent it from being bent or broken, the base was located in a "swamped" dip, so that the leaf lay flat on the receiver, see Figure 13-10. On the later 12,000 Type II rifles, the "swamped" dip was omitted; the rear sight was moved to 3-1/4 inches ahead of the receiver and mounted on the top surface of the barrel. A small screw, or pin, with a slotted head was inserted into the top, underside of the sight leaf to rest against the barrel. This provided the necessary support to the leaf to prevent breakage, see Figure 13-11.

Bayonet

Unlike any other Springfield arm of this period, but in keeping with previous Navy tradition, this rifle mounted a sword bayonet rather than the angular bayonet. Most such bayonets were mounted on the right side of the barrel, but in the Model 1870 Navy rifles, they were mounted beneath the barrel. See Chapter 20 for a complete description of the Navy bayonets.

Ramrod

The U.S. Model 1868 Springfield-designed ramrod with the single shoulder was used with the U.S. Model 1870 Remington Navy rifle. It was 29-1/2 inches long and was intended to be used to clean the bore from both ends. The slotted head was 0.425 inch in diameter and had seven rings. The face was cupped. The ramrod was 1-11/16 inches short of the muzzle when stowed, see Figure 13-12. The ramrod entered a hole in the bayonet hilt when the bayonet was mounted.

Fig. 13-7. The Navy inspector's initials and the firing proof mark were stamped on the left side of the U.S. Model 1870 Type I Navy rifle barrel. Compare them to the more neatly stamped markings on the Type II rifle in Fig. 13-8.

Fig. 13-8. Navy inspector's marks photographed on a U.S. Model 1870 Type II Navy rifle.

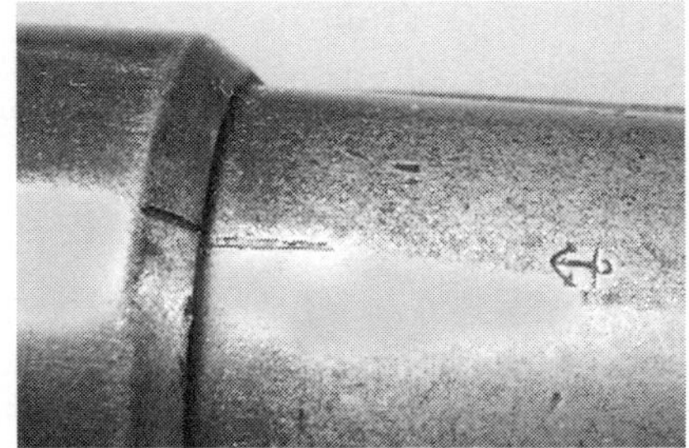

Fig. 13-9. The Navy's acceptance "anchor" mark was stamped on the side of the barrel adjacent to the barrel and receiver alignment marks.

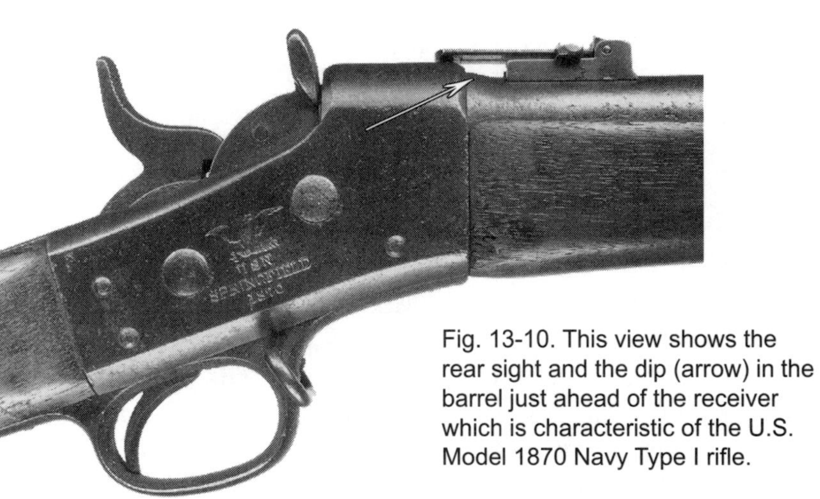

Fig. 13-10. This view shows the rear sight and the dip (arrow) in the barrel just ahead of the receiver which is characteristic of the U.S. Model 1870 Navy Type I rifle.

Fig. 13-11. This view shows the rear sight on the U.S. Model 1870 Navy Type II rifle moved forward on the barrel, the elimination of the dip ahead of the receiver, and also the pin (arrow) in the top of the sight leaf that supports it when laid flat.

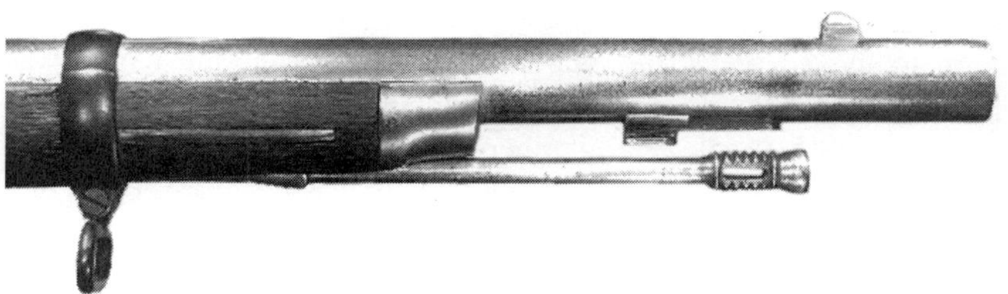

Fig. 13-12. This view of the forend and muzzle shows the ramrod setback and side view of bayonet lug. Note the blued barrel band.

Variant in .22 Caliber

In 1889, one hundred (or possibly less) of these arms were converted to .22 rimfire by the Winchester Repeating Arms Company for target practice aboard ship. This is the first known use of the .22-caliber cartridge by the U.S. government. Barrels were modified by the installation of a rifled liner, which (at only about 24 inches in length) stops well short of the muzzle, and therefore is not apparent when examining only the exterior of this extremely scarce arm, see Figure 13-13, arrow A.

Fig. 13-13. The barrel liner (A) and the altered extractor (B) for the .22-caliber cartridge are clearly visible in this photograph. Photo courtesy of Roy Marcot.

The extractor mechanism (arrow B) was altered to handle the smaller cartridge, and the firing pin was offset so as to strike its rim. In addition to the normal proof marks described above, the barrel was marked on the left side, just above the wood, with the inspector's initials and an additional serial number, "No. XX W.W.K. 1889." The initials are those of William W. Kimball, the Navy inspector. Only one- and two-digit numbers have been observed to date.

Chapter 14: U.S. Model Remington Rifle Musket, Experimental Model 1870 (Remington Patents)

Historical Background

The U.S. Model Remington Rolling Block Rifle Musket, Experimental Model 1870 (also known to collectors as the U.S. Model 1870 Army Trials Rifle) was produced in response to an order from the "St. Louis Ordnance Board on Tactics, Small-Arms, and Accoutrements" for a series of trials conducted in 1871 and 1872. The board was also known as the "Schofield Board" after its head, Major General John Schofield. The Remington rolling block design, along with the Allin, Sharps, and Ward-Burton, was manufactured for and tested under field conditions. Although the Remington system was the first choice of the board, the trials ultimately showed that the Allin "trapdoor-system" was the overwhelming favorite with the troops. Since the Ordnance Department concurred, the deck was well and truly stacked in favor of the "Allin" system, and it would have taken a miracle to produce a major change in rifle design in the foreseeable future.

NOTE: There were two variations of the Remington rolling block rifles manufactured for the U.S. Army: the Model Remington Rolling Block Rifle Musket, Experimental Model 1870 rifle and carbine described in this chapter and the U.S. Model 1871 Army (Remington) which was produced in much greater quantity during 1872 and is described in Chapter 15.

Quantity Produced

A total of 1,013 Model 1870 Remington rifles (Figure 14-1) and 313 carbines (Figures 14-2 and 14-3) with the rolling block action were manufactured at the Springfield Armory for the Schofield Board–ordered field trials in .50-70 caliber.

Adapted From

Like the foregoing Navy models, the Model 1870 Army was another variation of the Remington No. 1 action, manufactured under license, at the Springfield Armory. The Trials rifle was quite similar to the two Navy models, differing only slightly in barrel length and markings. The Army model also differed from the Navy model in that it was intended for use with the conventional M1855 angular bayonet, rather than the heavy brass-hilted sword bayonet used by the Navy, see Figure 14-4.

The Model 1870 rifles have the same barrel length as the "production" Model 1871 Army rifle which followed, but do not have the latter's complicated "locking" action. In a very real sense, they are an intermediate step between the 1870 Navy rifle, and the fully improved 1871 Army rifle. The Trials rifles are occasionally overlooked by collectors, as they appear very similar to the more common varieties of Remington rolling block rifle.

The Model 1870 carbines manufactured for the trials are extremely rare today, with less than ten specimens known to have survived. The author has not physically inspected a Springfield Armory example of either Rem-

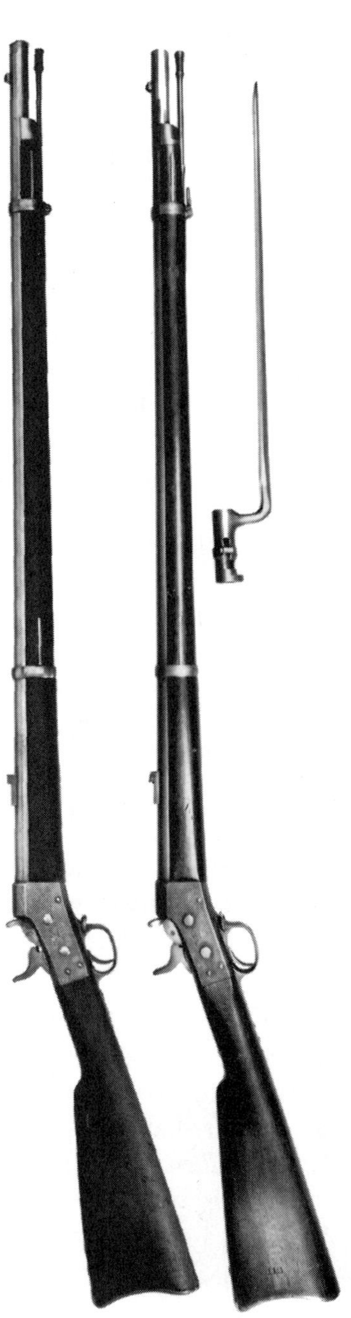

Fig. 14-1. Two examples of the U.S. Model 1870 Remington patent rifles manufactured for the 1871–1872 trials, ordered by the St. Louis Ordnance Board. Left, from the Smithsonian collection; right, from the West Point Military Academy collection.

Fig. 14-2. The U.S. Model 1870 Trials carbine, based on the Remington patented action, right side. Springfield Armory NHS collection, SPAR #6022. The rear sight is missing.

Fig. 14-3. The U.S. Model 1870 Trials carbine, based on the Remington patented action, left side. Springfield Armory NHS collection, SPAR #6007.

Fig. 14-4. The U.S. Model 1870 Trials rifle was fitted with a new U.S. Model 1868–type barrel. Accordingly, the Model 1855 bayonet was issued with the rifle.

ington 1870 trial arm, but was fortunate enough to have been able to consult with owners of each of them. The Springfield Armory Museum provided the photographs of the "sample" Army Trials carbine from their collection (SPAR #6007) through the courtesy of John McCabe of the National Parks Service and David Arnold, Conservator. The photograph of one of the Model 1870 rifles shown in Figure 14-1 was generously provided by the West Point Museum.

Other photographs and data were furnished by Roy Marcot, an authority on Remington Arms. The table and following descriptions are based on those consultations.

Table 14 U.S. Model Remington Rolling Block Rifle Musket, Experimental Model 1870	
Finishes	
Receiver	Color Case-Hardened (including trigger guard)
Breechblock or Bolt	National Armory Bright
Hammer	National Armory Bright
Barrel	National Armory Bright
Furniture	All National Armory Bright unless otherwise noted Rear sight: Browned (blued)
Stock	Oil-finished American Black Walnut
Markings	
Receiver	Upper tang: Remington patent markings. Right side: Eagle clutching arrows and olive branch/ "U.S./SPRINGFIELD/1870"
Breechblock or Bolt	Thumbpiece: knurled edge to edge
Hammer	Hammer: knurled edge to edge
Barrel	Witness mark at top rear center
Furniture	Butt Plate: "U.S." on tang Bands (2): "U" near upper edge, right side Rear Sight: (rifle) "2," "3," "5," "7," "9" right rear face of leaf (carbine) "2," "3," "5," "7" only on leaf (no 900-yard mark)
Stock	"ESA" in oval on left wrist. No date "BL" in box behind trigger plate
Principal Dimensions	
Overall Length	Rifle: 52 inches Carbine: 38 inches
Stock Length	Buttstock (Rifle and Carbine): 13-1/2 inches Forend, Rifle (including nose cap): 31-5/16 inches Band shoulders: 19-1/8 inches apart Forearm, Carbine (no nose cap): 11-3/4 inches

Barrel Length	Rifle: 36 inches; 3 grooves, 1 turn in 42 inches to the right Carbine: 22 inches; 3 grooves, 1 turn in 42 inches to the right
Muzzle Diameter	Rifle: 0.775 inch Carbine: 0.792 inch
Ramrod Length	Rifle: 34-1/14 inches (flush with the muzzle when stowed)

Identifying Features

Stock

Following the standard Remington rolling block pattern, the buttstock and forend or forearm were made in two separate pieces. The buttstock was 13-1/2 inches in length, measured along the bottom, from the toe to the lower corner of receiver, see Figure 14-5. It was inletted for the upper and lower tangs, and fitted for the Model 1863 "U.S."-marked rifle musket butt plate. The left wrist flat was stamped with the cartouche "ESA" in a small oval. A small boxed "B," "L," or "BL" was stamped behind the trigger guard.

Fig. 14-5. The U.S. Model 1870 Trials rifle and carbine used the same 13-1/2-inch-long buttstock. Photograph courtesy of Roy Marcot.

The rifle's forend was 31-5/16 inches long, including nose cap, but not the tenon, which fitted into the front of the receiver. It was held in place by the barrel-block. Two barrel bands, the upper of which carried a sling swivel, were used. Barrel band shoulders in the rifle forend were cut 19-1/8 inches apart, see Figure 14-6.

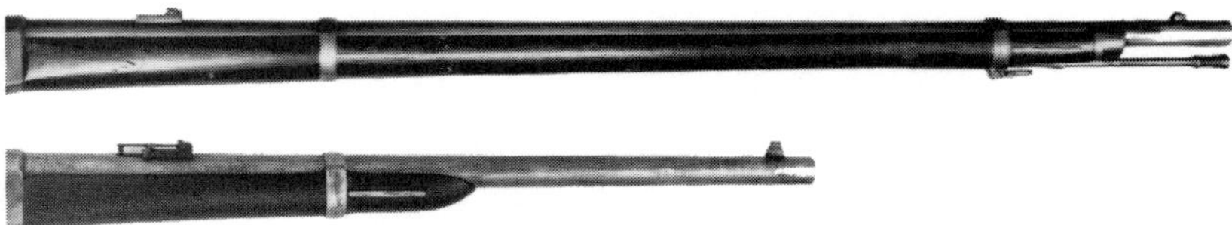

Fig. 14-6. The U.S. Model 1870 Trials rifle forend and carbine forearm are shown here for comparison. Photograph courtesy of Roy Marcot.

The carbine's forearm did not have a nose cap but tapered to a tip and was only 11-3/4 inches long overall, refer to Figure 14-6. The shoulder for the barrel band was cut 3-1/2 inches behind the forearm tip. The band spring was located on the right side, following Springfield's standard practice. This is a point of identification for the Model 1870 Remington Trials carbine as almost all other Remington-made, rolling block-pattern carbines have much shorter forearms with the band spring mounted on the bottom.

Action

The receivers used for both the rifle and carbine were marked on the right side with a large spread-winged eagle clutching arrows, above "U.S." above "SPRINGFIELD" above "1870," see Figure

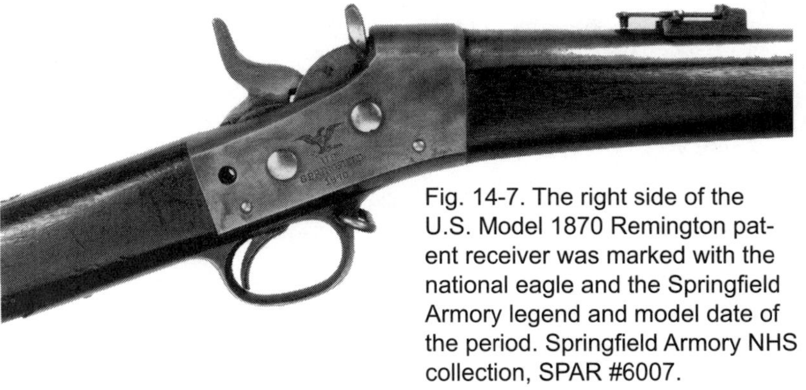

Fig. 14-7. The right side of the U.S. Model 1870 Remington patent receiver was marked with the national eagle and the Springfield Armory legend and model date of the period. Springfield Armory NHS collection, SPAR #6007.

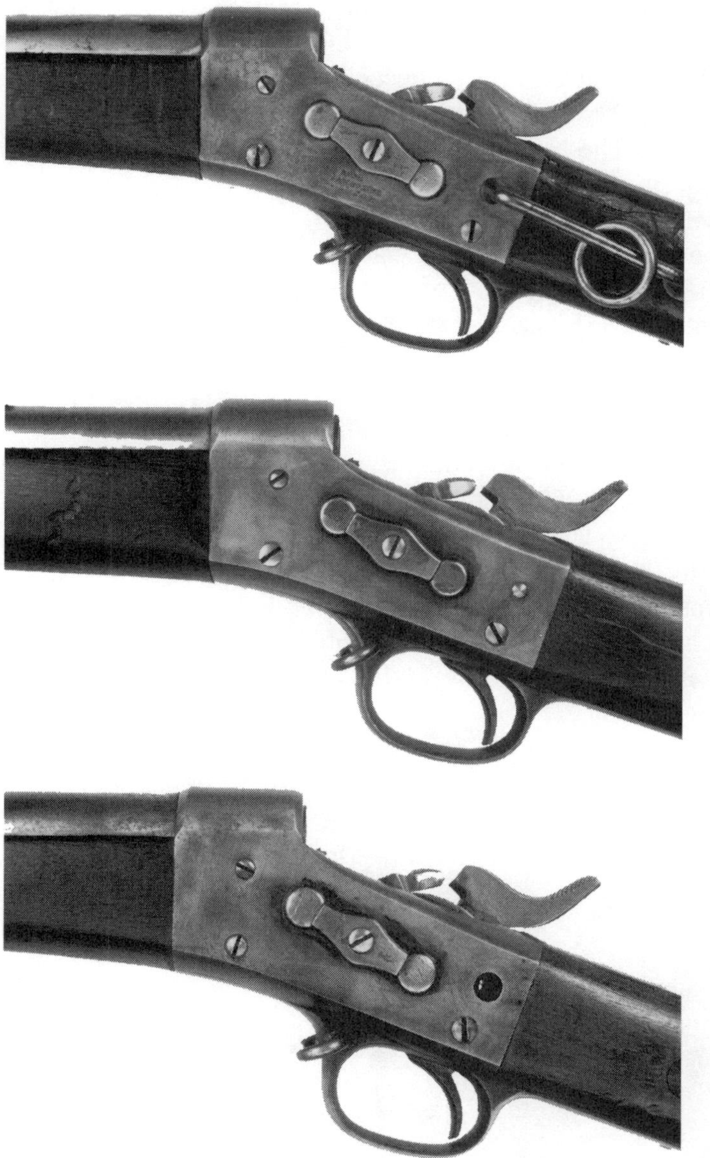

Fig. 14-8. The left side of the U.S. Model 1870 Remington patent receiver was unmarked. The only exception noted to date is carbine SPAR #6007 (top) in the Springfield Armory collection. Below, rifle SPAR #5995 and carbine SPAR #6022. The sling bar is missing on this carbine.

14-7. The left side was entirely unmarked except for one example in the Springfield Armory collection, SPAR #6007, as shown in Figure 14-8, top. The hammer profile was similar to that of the 1870 Navy models described in Chapter 13 and noticeably smaller than that of the 1871 Army model described in Chapter 15. The hammer spur was bent back slightly from the vertical. The knurled thumbpiece on the breechblock projected upward. Like its Navy predecessor, this action used a sliding extractor, see Figure 14-9, arrow. The lower sling swivel was mounted in the front swell of the massive one-piece trigger guard, which itself was an integral part of the lower tang, see Figure 14-10.

Fig. 14-9. The sliding extractor similar to that used in the U.S. Model 1870 Navy Rifle was also used in the U.S. Model 1870 Trials rifle. Photograph courtesy of Roy Marcot.

The U.S. Model 1870 Trials rifle trigger guard, with its integral swivel, was also used on the U.S. Model 1870 Trials carbine, even though a rifle sling was not employed by mounted troops, refer to Figure 14-10. It may have been decided that making up a different trigger guard for less than 350 test carbines was a waste of time and money.

Barrel

The length of the rifle barrel was 36 inches; otherwise it was exactly the same pattern as used on the second

Fig. 14-10. The trigger guard with its integral sling swivel was used on both the rifle and carbine. Although a rifle sling was not used with the carbine, it was retained in the interest of economy. Photograph courtesy of Roy Marcot.

type U.S. Model 1870 Navy rifles. It was fitted with the Springfield one-piece combination front sight and bayonet lug. The barrel was tapered to 0.775 inch in diameter at the muzzle to accept the standard Model 1855 socket bayonet.

The carbine barrel was 22 inches long, and measured 0.792 inch in diameter at the muzzle. The specimen reported upon was marked "22," stamped on the bottom of the barrel, near the receiver.

Rear Sights

The U.S. Model 1868 rear sight was located 3-1/8 inches forward of the receiver on both the Model 1870 Trials rifle and carbine. The rear sight was held in a dovetail mortise and by one spanner-head screw. As in the other trial arms, the rifle's elevation slide was 0.245 inch in height; that used on the carbine was 0.169 inch. A pin, 5/32 inch (0.16 inch) long, was screwed to the top rear edge of the leaf to prevent its being damaged when it was folded down against the barrel, see Figure 14-11.

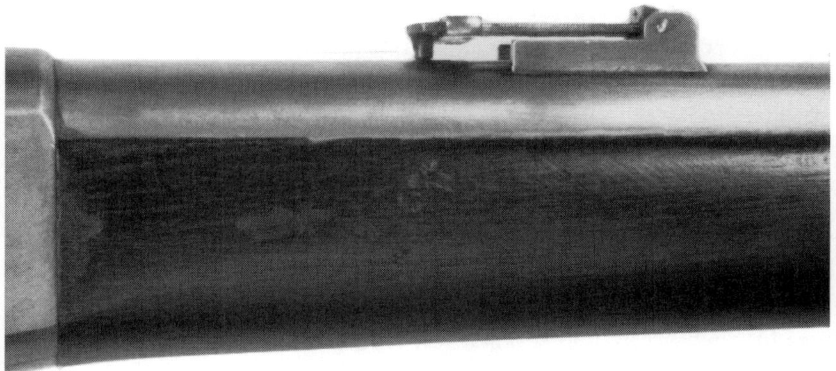

Fig. 14-11. The U.S. Model 1868 rear sight was used on the U.S. Model 1870 Trials rifle and carbine. Note the pin that supported the leaf when in the down position. Photograph courtesy of Roy Marcot.

Furniture

The U.S. Model 1863 rifle musket butt plate, marked "U.S.," was used on both the rifle and the carbine.

Two barrel bands were installed on the rifle. They were taken from surplus U.S. Model 1863 rifle muskets and both bands were marked "U" at upper right edge. Surplus U.S. Model 1863 rifle musket nose caps were also used on the rifle forend. They were unmarked.

The single carbine band used was similar to the thin-walled type used on the U.S. Model 1869 Cadet Rifle and the U.S. Model 1870 carbine, refer to Figure 9-12 in Chapter 9.

A sling ring bar was used as an attaching point for the mounted trooper's cavalry sling, see Figure 14-12. It was 2.85 inches long and ran from the left wall of the receiver to a plate 0.50 inch long by 0.923 inch high mortised into the left wrist. An unusual feature on the carbine was the retention of the rifle lower swivel for reasons of economy, mounted to the front of the trigger guard, even though it was not intended to be used as such.

Fig. 14-12. A sling bar and ring were mounted on the left side of the U.S. Model 1870 Trials carbine. Springfield Armory NHS collection, SPAR #6007.

Ramrod

The U.S. Model 1868 ramrod with the single shoulder and mortised ramrod stop was used on the U.S. Model 1870 Army Trials Rifle. The rod was 34-1/4 inches long and flush with the muzzle when stowed. It was finished bright and had a seven-ringed uncupped head. The shank lacked both thread or cannelures, see Figure 14-13. The carbine was not equipped with a ramrod.

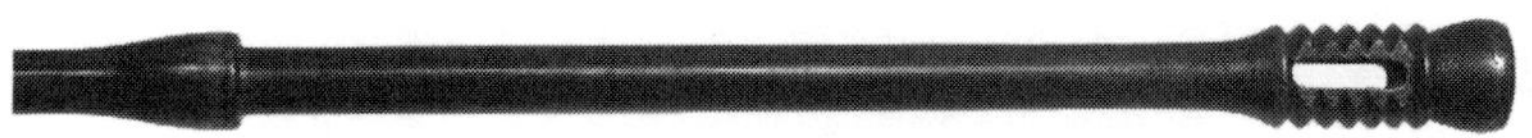

Fig. 14-13. The U.S. Model 1868 ramrod with the single shoulder was used with the U.S. Model 1870 Remington Trials rifle.

Chapter 15: The U.S. Model 1871 Army Rifle (Remington Patents)

Historical Background

The U.S. Model 1871 Remington Army Rifle was the last long arm equipped with the Remington system rolling block action manufactured by the U.S. military establishment.

Based on the quantity produced, 10,000, it certainly seemed that some parties hoped this "improved" version of the Remington system would be adopted in place of the Allin system. But it was not to be.

The Army's field trials of 1871–1872 (requested by the Schofield Board meeting at St. Louis) had finally resulted in the selection of the Allin system, to the exclusion of all others. This decision was confirmed by the Boards on Breech Loading Small-Arms and Calibre (Terry Board), sitting at New York in 1872–1873, after testing no less than 99 contenders, including some from abroad.

Concurrently, Board for the Purpose of Determining the Calibre of Small-Arms had been convened at the Springfield Armory and charged with selecting a cartridge in a smaller caliber for the service arm. Experiments were made with cartridges of .45, .42, and .40 caliber. This board kept in close contact with the Terry Board; the final "joint" recommendation was the arm listed on the docket as "Springfield No. 99," which became the new service rifle, the "U.S. Model 1873 .45-70 Springfield Rifle."

Fig. 15-1. U.S. Model 1871 Remington Rifle in .50-70 caliber with the Remington patent breech. This was the last, and by far the most numerous, of the Springfield-manufactured Remington rolling block rifles.

The .58- and .50-Caliber Rifles

The U.S. Model 1871 Remington Army rifles were used by various state National Guard and militia units, but were never issued to the regular Army. Apparently, the funds to manufacture the rifles were drawn from the appropriations for arming the Militia. It should be noted that the U.S. Model 1871 Remington-system rifle in .50-70 caliber used in the trials was not the same rifle as the so-called "New York State" Remington-system rifle with its very tall hammer spur. That particular rifle (identifiable by its three barrel bands) was produced entirely by Remington Arms in their Ilion, New York, factory under contract to New York state.

Quantity Produced

A total of 10,000 U.S. Model 1871 Remington Army Rifles were manufactured at the Springfield Armory in 1872, see Figure 15-1. A carbine version of this model was not produced. The rifles were not serial numbered. The U.S. Model 1871 Remington Army Rifle is by far the most often encountered Springfield Armory–manufactured rolling block model.

Adapted From

The U.S. Model 1871 Remington Army Rifle was a further refinement of the two Springfield Armory–manufactured rolling block Navy rifles, as well as the two limited-production Army Trials models which preceded it. The various differences are detailed below.

Table 15 The U.S. Model 1871 Army Rifle (Remington Patents)	
Finishes	
Receiver	Color Case-Hardened (including trigger guard)
Breechblock or Bolt	National Armory Bright
Hammer	National Armory Bright
Barrel	National Armory Bright
Furniture	All National Armory Bright unless otherwise noted Rear sight: Browned (blued)

Stock	Oil-finished American Black Walnut
Markings	
Receiver	Upper tang: Remington patent markings. Right side: Eagle clutching arrows and olive branch/ "U.S./SPRINGFIELD/1872" Left side (below retainer plate): "MODEL 1871"
Breechblock or Bolt	Thumbpiece: knurled edge to edge
Hammer	Hammer: knurled edge to edge
Barrel	Witness mark at top rear center
Furniture	Butt Plate: "U.S." on tang Bands (2): "U" near upper edge, right side Rear Sight: "2," "3," "5," "7," "9" right rear face of leaf
Stock	"ESA" in oval on left wrist. No date "BL" in box behind trigger plate
Principal Dimensions	
Overall Length	52 inches
Stock Length	Buttstock: 13-1/2 inches Forend (including nose cap): 31-5/16 inches Band shoulders: 19-1/8 inches apart
Barrel Length	36 inches; 3 grooves, 1 turn in 42 inches to the right
Muzzle Diameter	0.775 inch
Ramrod Length	34-1/4 inches (flush with the muzzle when stowed)

Identifying Features

Stock

The buttstock was 13-1/2 inches in length, measured from the toe to lower corner of the receiver. The buttstock was inletted for the upper and lower tangs, and fitted for the standard "U.S."-marked U.S. Model 1863 rifle musket butt plate. The left wrist flat was stamped with the "ESA" in a small oval cartouche, see Figure 15-2. The forend was 31-5/16 inches long, including nose cap, but not the tenon which fitted into the front of the receiver. The forend was retained by the barrel-block, and by two barrel bands. The upper barrel band

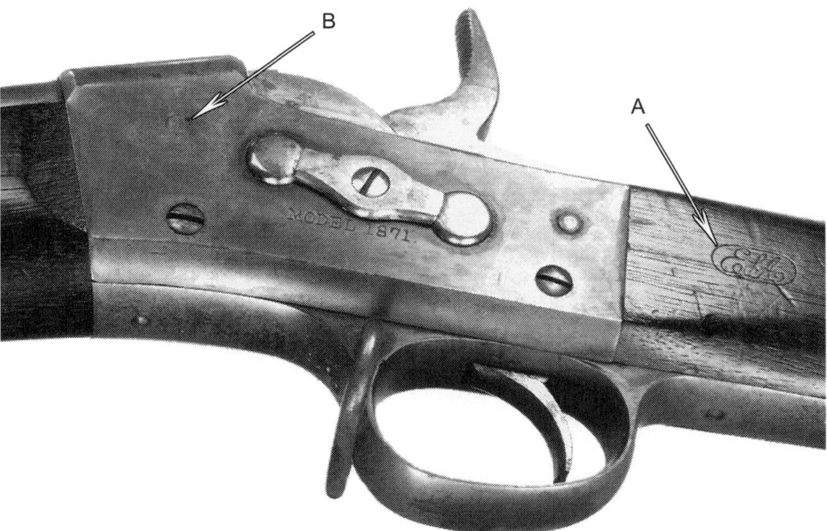

Fig. 15-2. The action of the U.S. Model 1871 Remington Rifle showing the markings on the left side of the receiver and the "ESA" inspector's stamp on left wrist (A). Also note the plugged hole note (B) for the screw which retained the sliding extractor which was used on the U.S. Model 1870 Remington Rifle (both Army and Navy models) but was replaced by a rotating extractor on this model.

was equipped with the forward sling swivel. The band shoulders were 19-1/8 inches apart.

Action

The Model 1871 Remington Army Rifle receiver was the most complicated form of the rolling block action yet designed. When the breechblock was opened, the cocked hammer was automatically released to rest against the breechblock, see Figure 15-3. It did not fall into any sort of half-cock notch, as other authorities have suggested; rather, it bore against the breechblock under pressure of the very strong mainspring, creating a noticeable drag as the block was closed. When the breechblock was pushed closed, the hammer *then* dropped into a half-cock position. To fire, the hammer had to be brought to full cock again. The system was intended to prevent the rifle from being

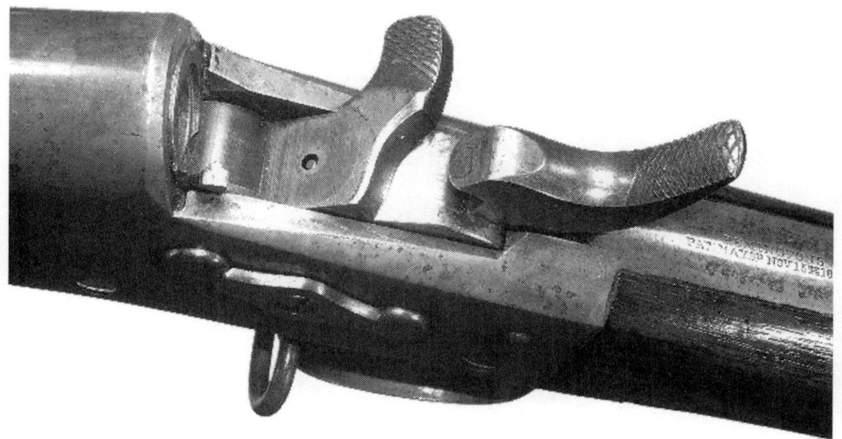

Fig. 15-3. With the action open, it can be seen that on the U.S. Model 1871 Remington Rifle, the breechblock thumbpiece extends out to the right, instead of upward as in the Model 1870. Here, the hammer has fallen against the breechblock, and must be re-cocked before firing.

prematurely fired during loading or after the breechblock was closed. In practice, it would have been a nuisance.

Fig. 15-4. This view shows the rotating extractor (arrow), the final—and best—of the three styles of rolling block extractors.

Unlike the previously described Remington-system models which used sliding extractors, the U.S. Model 1871 Remington "locking" action utilized a rotating extractor, see Figure 15-4, mounted in the left side of the breechblock on the same pin. Other differences from previous Springfield Armory–built rolling blocks include a noticeably taller, much more vertically oriented hammer spur. Also, the breechblock thumbpiece was shaped differently than previous models. It projected out to the right rather than upward, refer to Figures 15-2 and 15-3. One possible reason for this change: it was immediately distinguishable from the hammer, even in the dark, simply by feel.

Markings on the right side of the receiver were (see Figure 15-5):

Eagle
U.S.
SPRINGFIELD
1872

On the left side just below the pin retainer plate (refer to Figure 15-2):

MODEL 1871

On the upper tang (see Figure 15-6):

REMINGTONS PATENT
PAT. MAY 3^{D} NOV 15TH 1864 APRIL 17TH 1868

Barrel

The barrel was 36 inches long, with a 0.50-inch bore and chambered for the .50-70 cartridge. The barrel was manufactured at the Springfield Armory, based on that of the U.S. Model 1868 Rifle. It was rifled with three broad lands and grooves, with a twist of one turn in 42 inches.

The one-piece Springfield Armory front sight as used on the rifle musket was mounted 1-1/4 inches behind the muzzle. The U.S. Model 1868 rear sight was installed on the barrel, see Figure 15-7. The muzzle diameter was 0.775 inch so that the Model 1855 angular bayonet could be used.

Furniture

The butt plate used was the standard U.S. Model 1863 rifle musket pattern. It was marked “U.S.” on the tang.

Fig. 15-5. U.S. Model 1871 Remington Rifle right-side receiver marking.

Fig. 15-6. U.S. Model 1871 Remington Rifle tang marking.

Fig. 15-7. The U.S. Model 1868 rear sight was installed on the U.S. Model 1871 Remington Rifle.

As in the U.S. Model 1870 Remington Trials Rifle, two barrel bands were used to secure the forend to the barrel. Both bands were marked "U" to indicate "up" or toward the front of the barrel. The nose caps were taken from surplus U.S. Model 1863 rifle muskets and were not marked.

Ramrod

The U.S. Model 1871 Remington Rifle was unique in the series of Springfield Armory-produced rolling blocks in that it used the improved 1872 pattern (or late U.S. Model 1870 Rifle ramrod) with double shoulders to prevent battering the muzzle during cleaning, see Figure 15-8. The rod was 34-1/4 inches long and when stowed, was only 1/16 inch (0.06 inch) short of muzzle, which, in the author's opinion, is "flush" with the muzzle. It was polished bright and the flat-faced head had seven rings. The shank was plain without threads or cannelures.

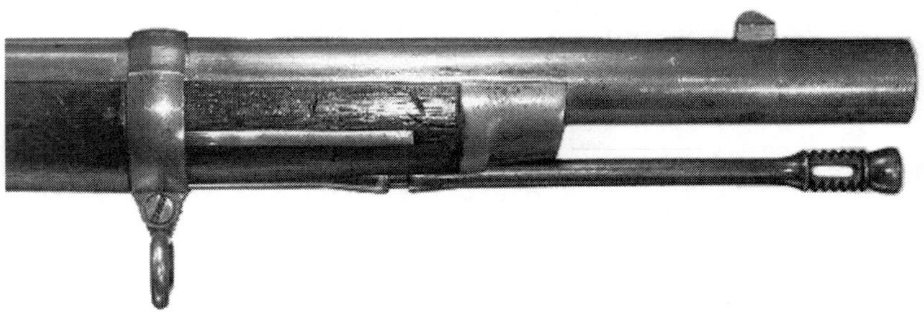

Fig. 15-8. This view shows the Model 1863 rifle musket upper band and the Model 1870 ramrod with the double shoulder. Of all of the 1871–72 Trials Rifles, only the U.S. Model 1871 Remington Rifle, as manufactured at the Springfield Armory, was equipped with the double-shoulder ramrod.

Chapter 16: The U.S. Model 1871 "Infantry Rifle" (As Altered from the Model 1865 Spencer Carbine)

Historical Background

The period between 1865, the end of the Civil War, and 1873, when the .45-70 cartridge was standardized, was truly a time of flux in U.S. military small-arms design. Arms manufacturers desperately trying to stay in business after the war were jockeying for what would today be called "market share." The muzzleloading rifle was clearly dead and the self-contained metallic cartridge was becoming universally accepted, so it was no wonder that so many new ideas were being presented to the military for testing. Unfortunately, Congress was reluctant to spend money on new rifles when hundreds of thousands of surplus rifle muskets were in storage, even if they were obsolete. The Ordnance Department's solution was to disassemble those rifle muskets and use as many of their parts as possible in new rifles as a means of reducing costs.

The U.S. Model 1871 Springfield-Spencer Rifle was doomed from the start, even though it was a repeater, see Figure 16-1. In its converted form it was excessively heavy and had a complicated action. It fired an underpowered (by post–Civil War infantry rifle standards) cartridge with little range. In addition, it made small use of the stockpiled obsolete U.S. Model 1863 and 1864 rifle muskets and their parts. Once the limited stock of Spencer parts was exhausted, it would have been an expensive arm to maintain.

Fig. 16-1. The U.S. Model 1871 Infantry rifles were seven-shot repeating rifles altered from U.S. Model 1865 Spencer carbines.

Quantity Produced

Just 1,109 U.S. Model 1871 Springfield-Spencer Rifles were produced in the year 1871 and were intended for use by the U.S. Army and not sale abroad. All were made up from a group of cavalry carbines in the serial number range 1–34,500 that had been shipped to Springfield Armory from Ft. Leavenworth, Kansas, for repairs. Upon review by the Ordnance Department, it was decided to convert them to rifles instead. When the work was completed, they were placed in storage until 1876, when they were shipped to New Mexico for possible issue. But the converted Spencers were never issued and were later shipped once again to the Rock Island Arsenal in Illinois for further storage, and ultimately, disposal.

Adapted From

The parent arm was a Model 1865 Spencer seven-shot repeating carbine, in .56-.50 rimfire manufactured by the Burnside Rifle Company of Providence, Rhode Island, which had produced the arms in late 1864 and early 1865. All were equipped with the Stabler magazine cutoff located just ahead of the trigger. The Stabler cutoff limited the distance the finger lever could open the breechblock. Seven cartridges were carried in a magazine tube inside the buttstock. The half-opened breechblock prevented them from feeding into the chamber and allowed the soldier to insert one cartridge at a time into the exposed chamber. This saved the magazine for those times when rapid fire was needed. By turning the cutoff switch 90 degrees to the bore, the lever dropped all the way open and allowed cartridges to feed from the magazine.

It took nearly nine months to retool and set up the required machinery: the first Model 1865 carbines with the Stabler cutoff were not delivered until April 15, 1865, nine days after the last major battle of the Civil War at Saylor's Creek, Virginia, according to author Roy Marcot, an authority on the Spencer.

At the Springfield Armory, U.S. Model 1868 Rifle barrels were modified to fit the Spencer action, then browned (blued). A new forearm was manufactured (possibly from leftover Remington Navy blanks). The U.S. Model 1868 rear sight, two new barrel bands, a nose cap, and the U.S. Model 1868 single-shoulder ramrod completed the rifle. The U.S. Model 1871 Spencer Rifle was thus different in appearance to the original Spencer-built military rifles used during the Civil War. Instead of three barrel bands it had two; where the original lacked a ramrod, the Model 1871 was equipped with one stowed in a channel under the barrel. Where the original rifle was equipped with the single leaf rear sight with the exposed leaf spring, the new rifle used the U.S. Model 1868/1870 rear sight.

Table 16 **U.S. Model 1871 "Infantry Rifle"** **(As Altered from the Model 1865 Spencer Carbine)**	
Finishes	
Receiver	Color Case-Hardened
Breechblock or Bolt	Color Case-Hardened
Hammer	Color Case-Hardened
Barrel	Browned (blued)
Furniture	All National Armory Bright unless otherwise noted Rear sight, magazine cutoff: Browned (blued) Butt Plate/cover: Color Case-Hardened
Stock	Oil-finished American Black Walnut
Markings	
Receiver	Top flat: "MODEL/1865" (at right angle) and "SPENCER REPEATING RIFLE/ PAT' D MARCH 6, 1860/MANUF'D AT PROV. R.I./BY BURNSIDE RIFLE Co" Upper rear: Serial number (applied by Burnside)
Breechblock or Bolt	Single letter(s) on right side (concealed when action closed)
Hammer	Hammer: knurled edge to edge, no markings

Barrel	Witness mark at top rear center Left rear top: serial number (applied at Springfield Armory) should match receiver
Furniture	Bands (2): "U" near upper edge, right side Rear Sight: "2," "3," "5," "7," "9" right rear face of leaf
Stock	"ESA" in oval on left flat (sometimes two). No date May also show remnants of original cartouche(s)
Principal Dimensions	
Overall Length	49-3/4 inches
Stock Length	Buttstock: 14-1/16 inches (original Spencer component) Forend (including nose cap): 26-5/16 inches Band shoulders: 14-5/8 inches apart
Barrel Length	32-5/8 inches; 3 grooves, 1 turn in 42 inches to the right
Muzzle Diameter	0.775 inch
Ramrod Length	29-3/8 inches (1-1/2 inches short of muzzle when stowed)

Identifying Features

Action

The receiver housed the lever-actuated rotating breechblock, see Figure 16-2. The breechblock was manufactured in two sections: upper and lower, with a coil spring between, Figure 16-3. The receiver was quite narrow; at the front it was 1.185 inches wide, tapering to 1.015 inches wide near the center and then swelling to 1.618 inches wide at the rear to accept the buttstock and magazine tube. A serial number was stamped on the upper rear surface.

The top flat was marked in six lines on the top of the receiver (see Figure 16-4):

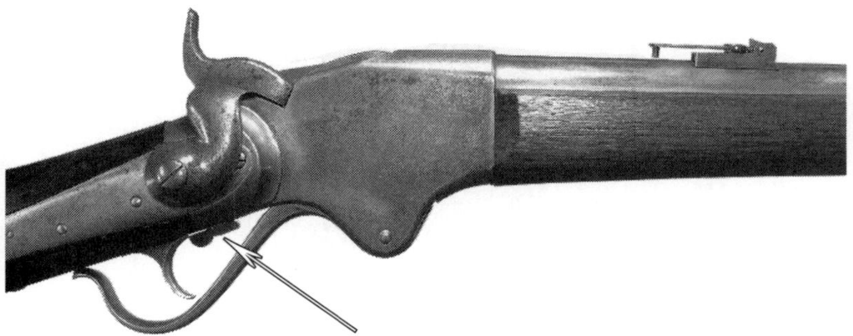

Fig. 16-2. The U.S. Model 1871 Spencer Infantry rifle used the Model 1865's back-action lock and Stabler cutoff (arrow) shown here in the "magazine off" position, which allowed the soldier to load one cartridge after each shot while holding the cartridges in the magazine in reserve.

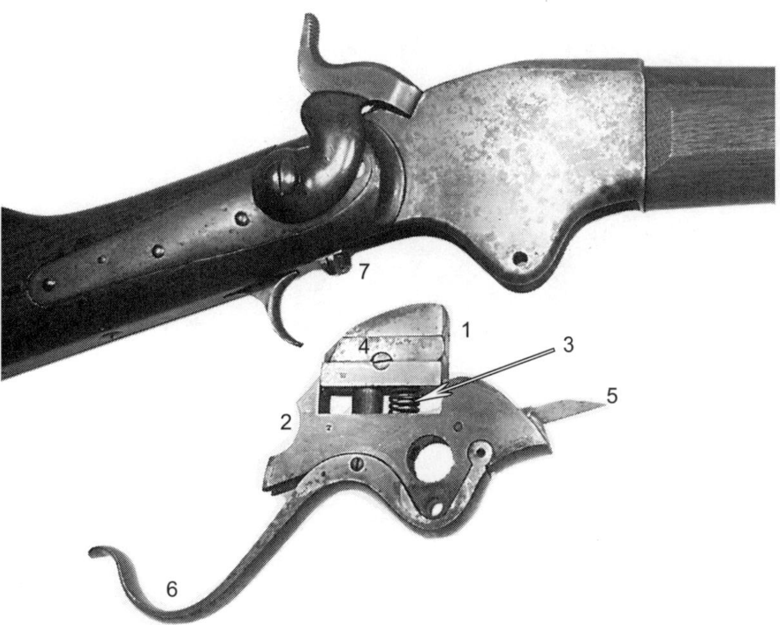

Fig. 16-3. The U.S. Model 1871 Infantry rifle with the Spencer breechblock and lever removed. The action consisted of 1) the upper and 2) lower sections of the breechblock, 3) the breechblock spring, 4) the firing pin with retaining screw, 5) the extractor, 6) finger lever, and 7) the Stabler cutoff. The cutoff is "off" (turned at right angle to the bore) which means cartridges can feed from the magazine. North Cape Publications collection.

MODEL
1865

at right angle to the bore, then parallel to the bore and reading from muzzle to breech:

SPENCER REPEATING RIFLE
PAT' D MARCH 6, 1860
MANUF'D AT PROV. R.I.
BY BURNSIDE RIFLE C°

Fig. 16-4. This view shows the model year, name, patent date and the Burnside Rifle Company name stamped on the top of the receiver.

When the new U.S. Model 1868 barrels were installed at the Springfield Armory, the serial number of the Spencer action was stamped on the barrel ahead of the receiver, on the left side, just above the forend, see Figure 16-5.

The Spencer was a fascinating piece of machinery. Upon opening the serpentine underlever, the upper section of the breechblock was pulled about 1/4 inch (0.25 inch) straight downward, against the pressure of a coil spring between the breechblock halves. This same

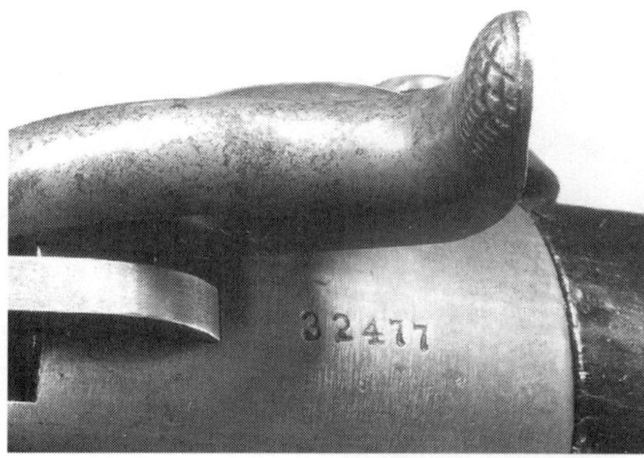

Fig. 16-5. The Springfield Armory stamped the existing Spencer serial number from the receiver onto the left side of their new barrel, using the same dies as on the U.S. Model 1868 .50-70 Springfield Rifle.

spring also held the lever closed. Once disengaged from its locked position against the rear of the receiver, the breechblock assembly rotated down to allow cartridges, under pressure from the spring in the magazine, to feed into the chamber. When equipped with the Stabler cutoff, the breechblock could only be rotated far enough open to eject a fired case and allow the shooter to insert a new cartridge, see Figure 16-6.

Cartridge cases were extracted by a "hook" (see Figure 16-7, arrow A) machined into the rotating lower breechblock assisted by a thin "knife blade" arm pivoted to the lower left front of the block, which "swept" the left receiver wall. The case was guided out of the open breech by a spring-loaded arm which pivoted up and down at the upper rear of the receiver, see Figure 16-8. This arm acted both as an exit ramp for the empty case and a guide for the next cartridge entering from the magazine.

With the Stabler cutoff lever in the "magazine off" position (parallel to the bore), a fired cartridge was ejected as described above but a new cartridge entering from the magazine was blocked by the upper half of the breechblock, see Figure 16-9, arrow. This allowed the shooter to manually load a new cartridge into the exposed chamber.

The sharply curved hammer, see Figure 16-10, was externally mounted on an unmarked, teardrop-shaped lock plate recessed into both the stock wrist and the rear of the receiver. The lock was of the "back-action" type, in which the mainspring was located behind the hammer. Although the Spencer was a repeating rifle, the hammer had to be independently cocked for each shot, as it was not actuated by the lever. The hammer nose was beveled on the side and appeared to strike the receiver, but in actuality, struck a small flat firing pin recessed into the upper breechblock.

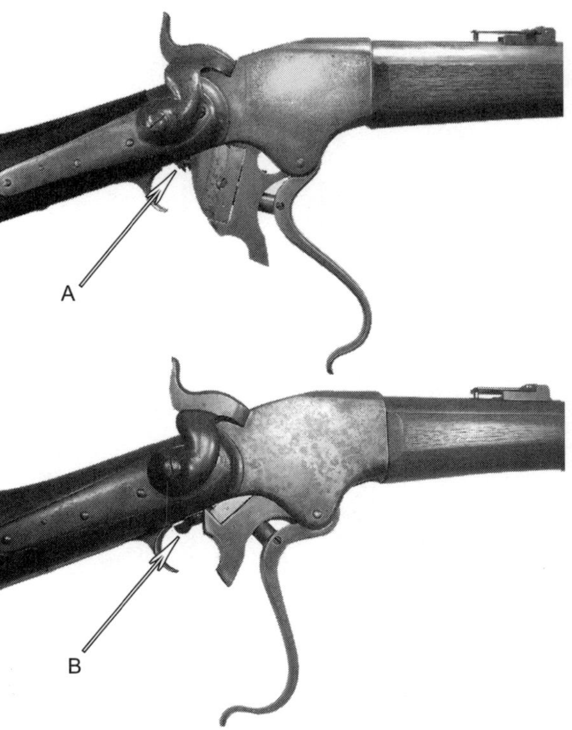

Fig. 16-6. Above: with the Stabler cutoff (A) turned 90 degrees to the bore, the finger lever lowers the breechblock sufficiently to allow cartridges to feed from the magazine in the stock. Below: with the cutoff "on" or turned in line with the bore (B) as in the lower view, the lever lowers the breechblock only enough to allow one cartridge to be manually inserted.

Magazine

The buttstock was bored through for a tubular magazine which held seven cartridges. As the cartridges used a rimfire priming system with the percussion compound distributed around the rim, there was very little danger of premature explosion caused by the pointed nose of a bullet striking the rear of the next cartridge. The magazine tube was

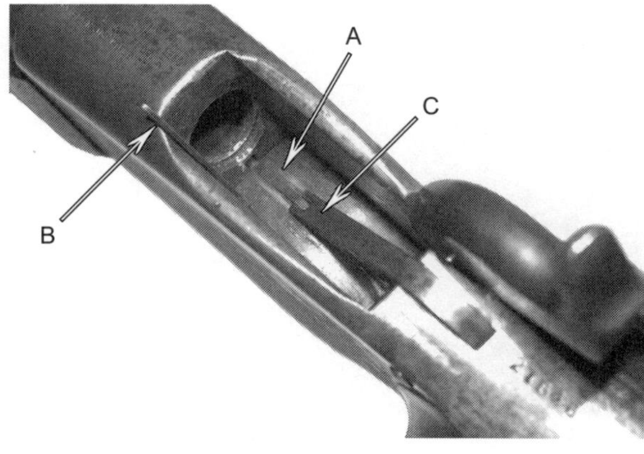

Fig. 16-7. This view of the partially opened action shows the extractor hook (arrow A) and the extractor proper (arrow B) which is pivoted at the bottom left side of the receiver. The cartridge guide (arrow C) is slotted to clear the extractor hook. North Cape Publications collection.

Fig. 16-8. Looking into a fully open action, the cartridge guide (arrow) is fully down, blocking any incoming cartridges. It now forms a ramp to guide the empty case up and out.

Fig. 16-9. A cartridge passes out of the magazine and is positioned by the cartridge guide (arrow) for insertion into the chamber as the finger lever is raised to close the breechblock.

held in place in the butt plate by an interrupted thread at the mouth of the tube, and a spring-loaded pin at the end of the cover arm.

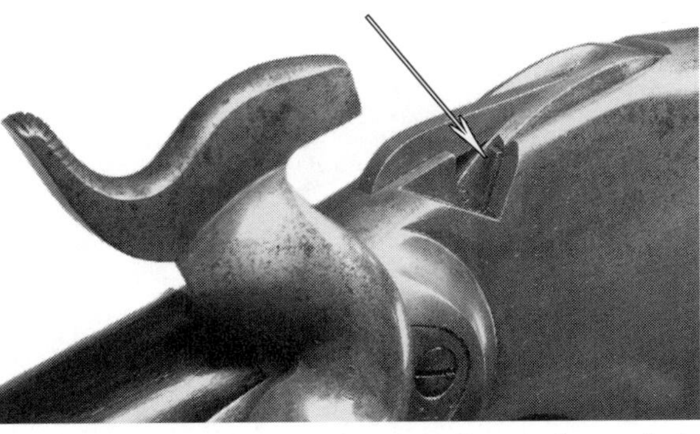

Fig. 16-10. With the hammer at half cock, the end of the firing pin is visible (arrow).

Loading was accomplished by twisting the magazine cover arm in the butt plate 90 degrees down and to the right, see Figure 16-11. The slotted inner magazine tube attached to the cover was drawn out, see Figure 16-12. Up to seven

Fig. 16-11. The original Spencer carbine butt plate, shown here with magazine cover closed, was retained on the converted arm.

Fig. 16-12. The loading cover has been twisted open and partly withdrawn to show the inner tube, and the coil spring which pushes the cartridges out of the magazine.

cartridges could then be inserted into the fixed or outer magazine tube, or liner, which remained in the stock. The inner tube was then replaced (telescoping over the cartridges) and locked in position. The cartridges were pushed forward by a spring-loaded follower located in the inner tube. One of the advantages of the Spencer magazine system was that it could be "topped off" while keeping one shot available, as the charging procedure did not require opening the breech.

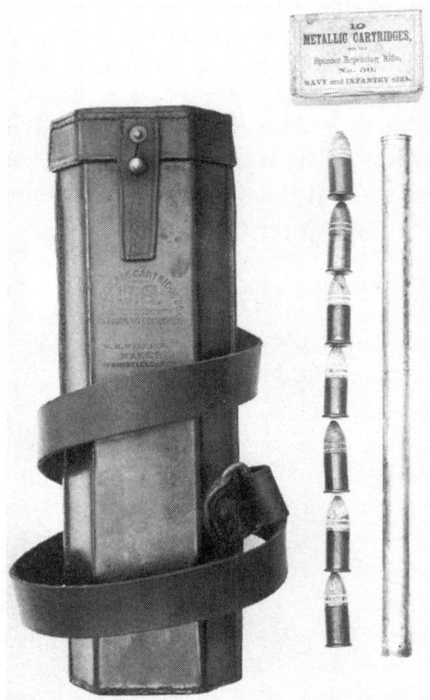

Fig. 16-13. This "cavalry model" Blakeslee Quick Loader cartridge box held ten tin tubes of seven cartridges, ready to pour into magazine. While used with the Model 1865 Spencer carbines issued to mounted troops after the end of the Civil War, it is not known if they were intended for issue with the U.S. Model 1871 Spencer rifle. Photo courtesy of Roy Marcot.

The Blakeslee Quick Loader was a leather-covered rectangular or hexagonal wooden block, bored with six, ten, or thirteen holes, closed with a hinged lid. It was worn slung on a strap across the soldier's chest or fastened to his waist belt. Each hole held a tin tube containing seven rounds, ready for quick insertion into the rifle, see Figure 16-13.

The "trap" in the butt plate of the U.S. Model 1877 and later carbines and the U.S. Model 1880, 1884 and 1888 ramrod bayonet rifles were unquestionably based on the Spencer design. The Spencer magazine principle lives on in an uncountable number of .22 rifles manufactured to the present day with nearly identical tubular buttstock magazines. Even the tube-charging concept survived until very recently, as the most convenient way of quickly reloading .22 rimfire rifles at carnivals and shooting galleries.

The .58- and .50-Caliber Rifles

Barrel

Barrels originally designed for the U.S. Model 1868 rifle were modified and installed on the U.S. Model 1871 Spencer rifles, but were chambered for the .56-.50-caliber Spencer rimfire cartridge. They were rifled with one turn in 42 inches in a right-hand twist. The rifling consisted of three broad lands and grooves. Unlike most other Springfield-produced rifles of the period, these barrels were blued instead of being given the usual National Armory Bright finish. The receiver serial number was stamped on the left side of the barrel, immediately ahead of the receiver and above the forend, refer to Figure 16-5.

The barrel length through the bore was 32-5/8 inches. The converted rifles were equipped with the standard one-piece combination front sight and bayonet lug. The barrel was tapered to 0.775 inch at the muzzle to accept the standard M1855 bayonet.

Rear Sight

The U.S. Model 1868 rear sight was mounted on the barrel 3-1/8 inches ahead of the receiver, see Figure 16-14. A pin 7/32 inch (0.22 inch) long was threaded into the top, rear side of the leaf to prevent damage when folded down.

Fig. 16-14. The U.S. Model 1868 rear sight with the two "V"-shaped sighting notches (arrows) was fitted to the Spencer conversions.

On all specimens viewed, the leaf gradations were "2," "3," "5," "7," and "9." However, some observers have reported other range values. This would make sense, as the ballistics of the .56-.50 Spencer rimfire cartridge were definitely inferior to the more powerful .50-70 round for which this sight was originally calibrated.

Stock and Forend

Unlike the Sharps conversions (Chapter 17) for which new buttstocks were made to accept the standard U.S. Model 1863 Springfield rifle musket butt plate, the U.S. Model 1871 Spencer Rifles retained their original Spencer buttstocks. This was doubtless due to the cost and impracticality of boring new stocks for the tubular magazine and fitting the Model 1863 butt plate for the magazine assembly, see Figure 16-15. The buttstock was fitted with a gun sling swivel, 1.70 inches wide by 0.55 inch high, located 3 inches forward of the toe. The buttstocks may show three cartouches with the last two belonging to Erskine S. Allin ("ESA" in an oval cartouche), stamped at a right angle

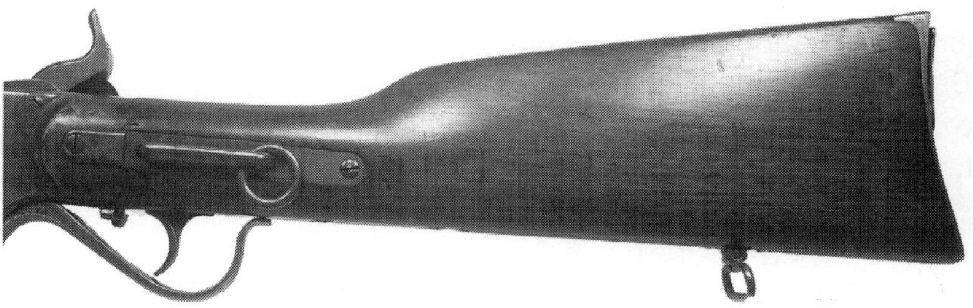

Fig. 16-15. The cavalry sling swivel was retained on the left side of the original Spencer carbine stock. North Cape Publications collection.

to the receiver, on the left wrist, behind the swivel bar plate. The third cartouche was originally applied by the Ordnance Department inspectors stationed at the Burnside Rifle Company during manufacture of the original carbines in 1865. One of three sets of initials in an oval cartouche will be found between the two "ESA" vertical cartouches: "HEV," "GC," or "LH," see Figure 16-16.

The forend was manufactured by Springfield Armory. It was 26-5/16 inches long, which left a longer-than-usual length (4-5/8 inches) of barrel exposed, see Figure 16-17. This was longer than any other Springfield arm of the period, except the U.S. Model 1870 Remington Navy rifles, which needed the extra space for the bayonet lug for the sword bayonet. In fact, from side-by-side examination of the rifles, it appears almost certain that leftover roughed-out Navy forend blanks were used to complete the relatively small run of converted Spencers, especially considering that they were produced as an afterthought.

Fig. 16-16. Multiple cartouches will appear on some but not all Springfield-Spencer conversions. Some of the specimens may show only the "ESA" stamp. North Cape Publications collection.

The new forend was machined for two barrel bands with shoulders 14-5/8 inches apart. The forend was equipped with a nose cap and was mortised to accept the U.S. Model 1868 ramrod. The ramrod retainer was inletted into the ramrod channel and held in place by the upper band spring pin.

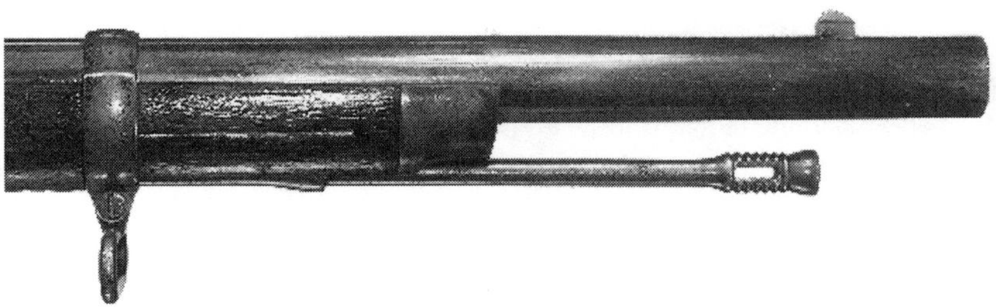

Fig. 16-17. This view of the muzzle end shows the unusually long setback of the forend and the ramrod, 1-1/2 inches short of the muzzle. It is thought that leftover Remington Model 1870 Navy blanks were used.

Furniture

Because the U.S. Model 1871 Spencer rifle was converted from a Model Spencer 1865 carbine, the altered model retained the sling bar and ring used by mounted troops, even though the rifles were intended to be issued to infantry. It is presumed that it was cheaper to leave the sling bar in place than to machine a new part to fill the mortise that would have been left when it was removed, refer to Figure 16-15.

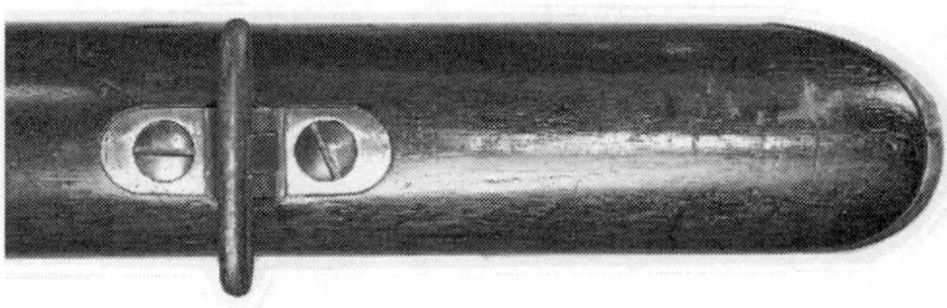

Fig. 16-18. The original Spencer butt swivel was retained. The same butt swivel was also used on the U.S. Model 1871 Sharps rifle and, much later, on the U.S. Model 1886 Experimental Carbine with the 24-inch barrel.

The parent Model 1865 Spencer carbine had been equipped with two conventional sling swivels. The forward sling swivel was mounted on the upper barrel band and the lower sling swivel on the buttstock 3 inches forward of the toe, see Figure 16-18. Because of the finger lever, the rear sling swivel could not be mounted in the preferred position on the trigger guard and so was inletted into the buttstock. This same lower sling swivel would show up again fifteen years later on the "Experimental" (XC) carbine of 1886.

Two new barrel bands were manufactured to the U.S. Springfield Model 1870 .50-70 rifle pattern. Both were marked "U" at upper right edge to indicate the side facing the muzzle. The forward sling swivel was mounted on the upper barrel band. The nose cap was similar to that used on other Springfield rifles of the period. It was 0.117 inch wide at the rear, tapering to 0.995 inch at the front and 0.94 inch long. As noted above, the original Spencer butt plate was reused to retain the trap and magazine assembly, refer to Figures 16-11 and 16-12.

Ramrod

Original Spencer rifles were not equipped with ramrods. The U.S. Model 1868 ramrod was added at the instructions of Major J.G. Benton, then commanding at the Springfield Armory. The ramrod was 29-3/8 inches long, had a seven-ringed head, was uncupped, finished in the white, and had a single shoulder. The shank was plain without threads or cannelures. When stowed, it was 1-1/2 inches short of the muzzle, refer to Figure 16-17.

Chapter 17: U.S. Model Sharps Rifle Musket, Experimental Model, 1870, Types I & II; Carbine, Type I Only

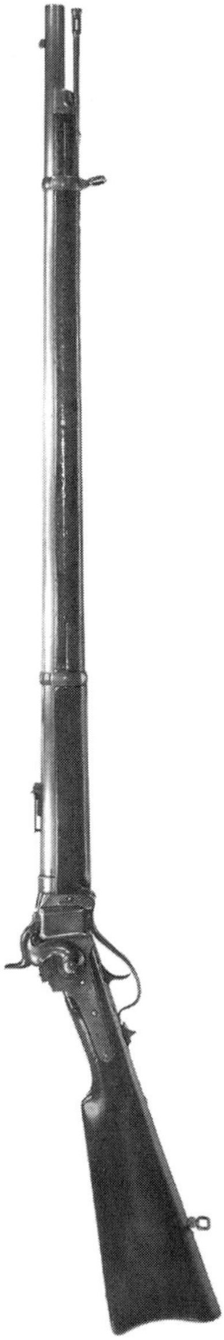

Historical Background

The Sharps breechloading system was first patented in 1848 and had proven itself as one of the most successful breechloaders of the Civil War. There had been problems in the early models with gas leakage from its linen-cased cartridge, but that issue had largely been overcome with the adoption of the Conant gas seal beginning with the Sharps Model of 1852. The Conant gas seal was an ingenious "blow-forward" ring in the breechblock face. The seal itself was a ring of platinum which the jeweler, A. Pegrot of New York City, furnished to the company at 42 cents each!

The Sharps breechloading action was immensely strong, and was easily adaptable to the longest, and most powerful cartridges, which the Spencer mechanism definitely was not. The action of the Sharps, like that of the Spencer, only opened and closed the breechblock and did not cock the hammer. That still had to be done manually for each shot.

The Sharps rifle and carbine were well liked by the troops during the Civil War; the federal government purchased 9,141 rifles and 80,512 carbines between 1861 and 1865 when the war ended, according to "Records of the Chief of Ordnance, Purchases 1861–1865." At the end of the war the Ordnance Department found that they had over 50,000 sur-

Fig. 17-1. U.S. Model 1870 Rifle with the Sharps action. Shown here is the Type I, serial #C37556, which was converted from a percussion "New Model 1863" action.

plus Sharps rifles and carbines, as well as spare parts, in inventory. The Sharps, as well as the Remington system, was considered, by those unhappy with the "trapdoor" design, to be a definite contender to unseat the "Allin-system" rifles as the preferred design.

As early as 1858, the Sharps Rifle Manufacturing Company had begun to work on converting the Sharps action to fire self-contained metallic cartridges. But the Ordnance Department, after some initial testing, declined to continue the work, as they were having enough trouble procuring approved models. But as the war was ending, the Ordnance Department again expressed an interest in metallic conversions. In February 1865, the Chief of Ordnance, General Dyer, requested the company furnish a price estimate for converting five thousand of the Sharps percussion carbines on hand and furnish suitable metallic cartridges for them. Competitive trials and preliminary work delayed the delivery of the first carbines until February 1868. By the end of 1869, a total of 31,184 Sharps had been converted to fire the .50-70 cartridge, the majority of them carbines. With this background, we can now discuss the relatively scarce Springfield-converted Sharps arms produced in 1871 which are somewhat different than their commercially altered precursors.

Quantity Produced

The Springfield Sharps was produced in two rather similar versions. The U.S. Model 1870 Sharps, Type I included 700 rifles and 300 carbines and was made using reworked surplus Sharps percussion actions salvaged from Civil War–era rifles, see Figure 17-1. These actions retained the mounting on the top edge of the lock plate for the Lawrence Pellet Primer system, although the internal parts were removed. The receivers were already serially numbered, and Springfield simply restamped the existing serial number (without the "C" prefix) from the Sharps action on the new barrel.

The U.S. Model 1870 Sharps, Type II included 300 rifles only. They were built on a new Sharps action specifically designed for metallic cartridges and manufactured by the Sharps Rifle Company. The actions for these Type II rifles came from a small lot of Sharps New

Model 1869 carbines which the Sharps company traded to Springfield for an equal number of Sharps percussion rifles. Springfield could not issue those rifles because they lacked spare parts.

No companion carbine was produced by the Springfield Armory with this style action. These rifles are serial numbered from 1–300, and production was completed in June 1871, see Figure 17-2.

Fig. 17-2. U.S. Model 1870 Rifle with the Sharps action; Type II, serial #226, a newly manufactured action designed for metallic cartridges.

Table 17 **U.S. Model Sharps Rifle Musket, Experimental Model, 1870,** **Types I & II; Carbine, Type I Only**	
Finishes	
Receiver	Color Case-Hardened
Breechblock or Bolt	Blackened* (Color Case-Hardened also known)
Hammer	Color Case-Hardened
Barrel	National Armory Bright
Furniture	All National Armory Bright unless otherwise noted Rear sight: Browned (blued)
Stock	Oil-finished American Black Walnut
Markings	
Receiver	Left side: "C. SHARPS' PAT./SEPT. $12^{\underline{TH}}$ 1848." Right side: None Upper tang: Serial number, applied by Sharps Type I (conversion): Circa C30,000-C40,000 Type II (new manufacture): Circa 1-300
Breechblock or Bolt	None
Hammer (all)	Hammer: knurled edge to edge, no markings

Lock Plate (Type I)	"C. SHARPS' PAT./OCT. 5TH 1852." and "R.S. LAWRENCE PAT/APRIL 12TH 1859."
Lock Plate (Type II)	None
Barrel (all)	Witness mark at top rear center Serial number at top left rear, just above wood (should match number on upper tang but without the "C" prefix) Type I (conversion): Circa 30,000-40,000 Type II (new manufacture): Circa 1-300
Furniture	Butt Plate: "U.S." Bands (2 for rifles, 1 for carbine): "U" near upper edge, right side Rear Sight (rifle): "2," "3," "5," "7," "9" right rear face of leaf Rear Sight (carbine): "2," "3," "5," "7" only; on leaf
Stock	"ESA" in oval on left wrist. No date "BL" in box behind trigger guard plate
Principal Dimensions	
Overall Length	Rifle (Types I and II): 52 inches Carbine: 38-15/16 inches
Stock Length	Buttstock (all): 15-1/2 inches along bottom edge Forend (rifle, including nose cap): 31-1/16 inches Forearm (carbine, without nose cap): 11-3/4 inches Band shoulders: 19-1/8 inches apart
Barrel Length	Rifle: 35-1/16 inches; 3 grooves, 1 turn in 42 inches to the right Carbine: 22 inches; 3 grooves, 1 turn in 42 inches to the right
Muzzle Diameter	Rifle: 0.775 inch Carbine: 0.792 inch
Ramrod Length	Rifle: 32-3/8 inches (flush with the muzzle when stowed)

* The term "blackened" is used to describe the color achieved by case-hardening in oil and water. The oil was floated on water. When the part to be case-hardened was removed from the oven, it was immediately immersed in the oil-water bath. The oil produced the black color and the water hardened the steel.

The .58- and .50-Caliber Rifles

Adapted From

The Type I rifles and carbines were made up from salvaged Sharps "New Model 1863" percussion actions, which included receiver and lock, and were fitted with new stocks, forends, furniture, breech-blocks, extractor systems, ramrods, and barrels; they were chambered for the .50-70 service cartridge. All parts but the actions were manufactured at the Springfield Armory.

The Type I carbines are *extremely* rare, with a survival rate of perhaps 10 or fewer. A specimen was not available for study. They appear, in profile, quite similar to the Sharps-produced Model 1867 and Model 1868 conversions. However, they had thicker buttstocks and U.S. Model 1863 rifle musket butt plates, notably longer forearms, and U.S. Model 1870 carbine rear sights, which were also installed on the U.S. Remington Model 1870 and U.S. Model M1871 Ward-Burton carbines.

The 300 Type II rifles were built in much the same way as the Type I rifles but using new actions manufactured by the Sharps Rifle Company. The actions were slimmer, simpler, and more streamlined and omitted the Lawrence Pellet Primer magazine. This very same action would ultimately become known in the Sharps commercial line as the "Model 1874" action.

Identifying Features

Buttstock

New buttstocks were manufactured at the Springfield Armory and were similar to those used on the various U.S. Remington system rolling blocks. They were inletted for the Sharps action's upper and lower action tangs, and were shaped to accept the U.S. Model 1863 rifle musket butt plate. The buttstock was marked "ESA" in a small oval cartouche on the left wrist, see Figure 17-3. The buttstock was equipped with a sling swivel taken from surplus Spencer parts. It was attached with two wood screws 2-7/8 inches ahead of the toe.

Forend

The forend was a new part manufactured at the Springfield Armory. It was 31-1/16 inches long, including the tip, 4-3/4 inches longer than the Model 1871 Spencer forend. The Model 1870 Sharps' barrel/muzzle configuration thus resembled that of the U.S. Model 1868 and Model 1870 Springfield Rifles, see Figure 17-4. The forend was mortised to accept a recoil block screwed to the barrel. Because of the way the receiver was made, the forend for the Sharps could not be formed with a tenon to fit into a matching recess in the front of the receiver.

Fig. 17-3. The patent marking stamped on the left side of the receiver was the same for both the Type I and Type II. Note the "ESA" cartouche on left wrist.

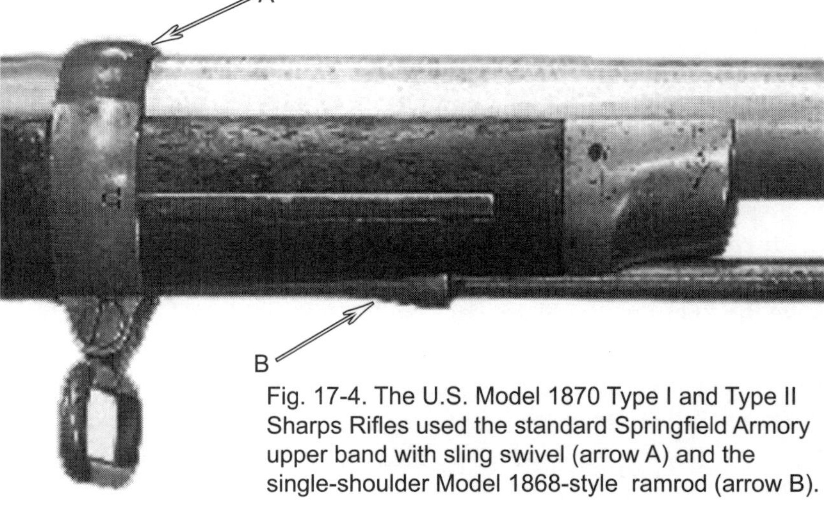

Fig. 17-4. The U.S. Model 1870 Type I and Type II Sharps Rifles used the standard Springfield Armory upper band with sling swivel (arrow A) and the single-shoulder Model 1868-style ramrod (arrow B).

Instead, it was fastened with a machine screw to the lever spring mounting block. Two Model 1870–pattern barrel bands (only one on the carbine) were used to secure the forend to the barrel. The barrel band shoulders were inletted into the forend and spaced 19-1/8 inches apart. The Model 1868 nose cap was also used.

At 11-3/4 inches, the uncapped carbine forearm was noticeably longer than the original Sharps item.

Action/Receiver—Type I

The receiver used on the Type I Model 1870 Sharps Rifle and Carbine was the "New Model 1863" percussion action, see Figure 17-5. It was distinguished by a long scroll-shaped lever which also served as trigger guard. The tip of the lever was retained by a spring-loaded sliding catch. The receiver had a vertical slot cut through from top to bottom for the massive breechblock, see Figure 17-6, arrow A. The breechblock used in the Type I was a new part manufactured by Sharps in which a firing pin was substituted for the cone of the parent arm. The breechblock was connected to the lever by a short link, see Figure 17-7, arrow, which pivoted on a screw in the right side of the block.

A long extractor, with a stubby arm extended at right angles, was pivoted at the bottom left side of the receiver on the lever/breechblock screw, see Figure 17-8, arrows A and B. When the finger lever was pulled down, the extractor snapped backward, drawing the case out of the breech, and, usually, all the way out of the receiver.

Like the Spencer, the Sharps used a "back-action" lock, in which the mainspring was located behind the hammer, refer to Figure 17-5. The trigger plate formed the lower tang and was fastened in a mortise in the bottom of the receiver with a single screw. The buttstock was held between the upper tang which formed an integral part of the receiver, and the lower tang, with a long machine screw that passed through the wrist, and two wood screws that entered from the bottom of the lower tang. The finger lever was held closed by a sliding latch in the lower tang.

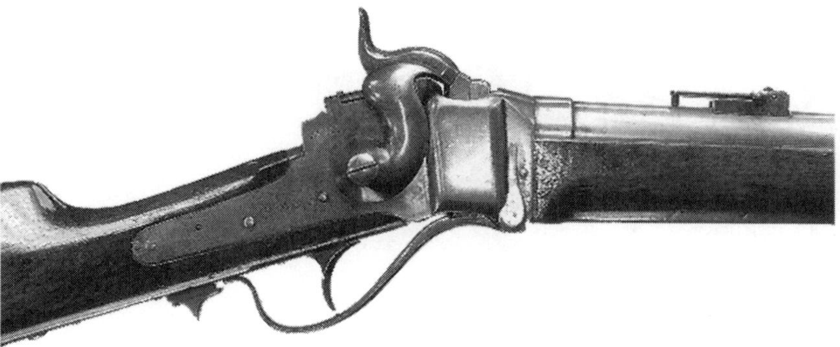

Fig. 17-5. This view of the U.S. Model 1870 Type I Sharps Rifle shows the vestiges of the Lawrence primer mechanism installed on the original percussion action. The interior parts were discarded during the conversion process.

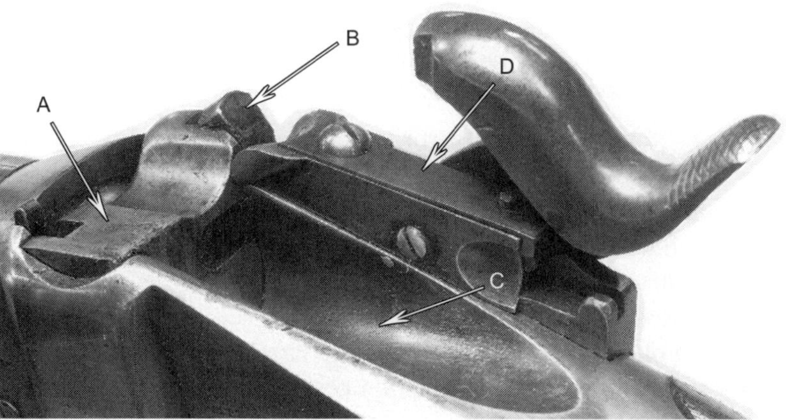

Fig. 17-6. This view shows the Type I breechblock (arrow A), the new firing pin (arrow B), the deep loading trough (arrow C), and the gutted Lawrence primer system (arrow D).

Fig. 17-7. In this view from beneath the action and with the lever open, the linkage (arrow) between the finger lever and the breechblock is visible.

Fig. 17-8. The long extractor (arrow A), pivoted on the breechblock. As the action was opened by lowering the breechblock, the extractor swung to the rear and its stubby cartridge-catching finger (arrow B) drew out the empty case.

The housing for the Lawrence disk primer magazine remained, although stripped of its internal parts. The half-cock notch for the hammer was noticeably higher than on other rifles and carbines equipped for the metallic cartridge. The hammer face was a half inch above the firing pin. This was a holdover from the time when it was configured as a percussion arm and the extra distance was needed for the soldier to be able to place a percussion cap on the nipple if the primer magazine was emptied, or failed to function correctly.

The receiver was marked in very small letters on the left side opposite the hammer, refer to Figure 17-3:

C. SHARPS' PAT./SEPT. 12TH 1848.

The lock plate was marked on the right side in two places—to the rear of the hammer:

C. SHARPS' PAT./OCT. 5TH 1852.

and, above the hammer, see Figure 17-9:

R.S. LAWRENCE PAT/APRIL 12TH 1859.

The latter patent referred to the Lawrence automatic primer feed. The internal mechanism of the primer feed was removed from the cartridge arm.

Fig. 17-9. The lock plates used on the U.S. Model 1870 Type 1 Sharps Rifles retained their original Lawrence and Sharps patent markings.

Action/Receiver—Type II

The Type II, Model 1870 Sharps action was based on the Type I, but was slimmer and less complicated, with a lowered and lightened hammer, see Figure 17-10. The lock plate was less than 0.19 inch thick, behind the hammer, as opposed to the 0.38 inch of its percussion predecessor. The tumbler was changed to provide a half cock just 0.152 inch above firing pin rather than 0.5 inch as in the percussion models used to build the U.S. Model 1870 Sharps, Type I rifles.

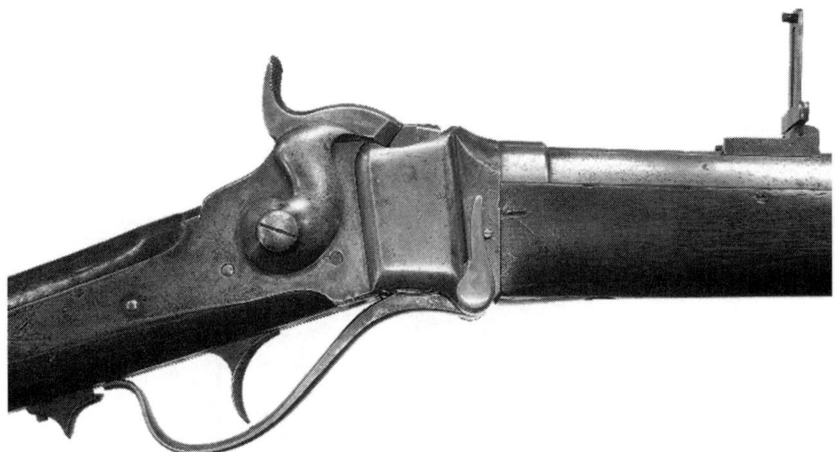

Fig. 17-10. The U.S. Model 1870 Type II Sharps Rifle action presents a cleaner configuration when compared to the Type I. The Sharps company would use this lock plate design on their commercial Model 1869 and 1874 rifles and carbines. Compare to Figures 17-5 and 17-12.

Contrary to descriptions in other sources, the friction roller remained in use on the lever spring. However, the small lug referred to above, at the front of the lever, was eliminated in the Type II action. Even so, no difference can be discerned in the "snap" of the extractor.

The marking on the left side of the receiver remained the same, but there were no markings on the right side of the lock plate.

Breechblock—Type I

New breechblocks were produced for all Sharps cartridge arms purchased or manufactured by the U.S. Army. The major change involved replacing the percussion nipple with a firing pin and eliminating the Conant gas seal and flash holes in the breechblock (referred to as the "slide" by the Sharps company). Since the breechblock was so compact in length (only 0.762 inch from front to rear) and the hammer was so far offset from the centerline of the bore, this new firing pin was basically a "U"-shaped sliding block, with one corner receiving the hammer blow and transmitting it to the opposite leg to strike the primer, see Figure 17-11. The firing pin was cammed back as the action was opened. The mechanical forces acting on this odd-shaped

Fig. 17-11. The new breechblock used in both the Type 1 and Type II rifles replaced the original percussion nipple with a complex firing pin. It also had a slot cut for the extractor needed to remove the empty metal cartridge case. The Lawrence primer system housing remained on the reused Type I lock plates.

firing pin caused a large number of them to break. And at best, they delivered a relatively weak blow to the primer, which resulted in an unacceptable level of misfires during the trials.

The Type I breechblock was 2.275 inches high, measured at the right side, by 1.490 inches wide. It was thus slightly taller (0.137 inch in that location) than the Type II breechblock.

Breechblock—Type II

The firing pin for the Type II breechblock was smaller and had a somewhat different shape than the Type I firing pin. The hammer blow was thus very slightly more in direct alignment. But the complicated shape of the firing pin continued to make it susceptible to breakage and it caused a significant number of misfires during the trials. The Type II block was slightly shorter, at 2.138 inches, on the right side only, than the Type I. The left side of both blocks was the same height, see Figure 17-12.

Barrel

The Sharps receiver required a larger-diameter barrel than the U.S. Model 1868/1870 barrel blanks on hand at Springfield. Rather than

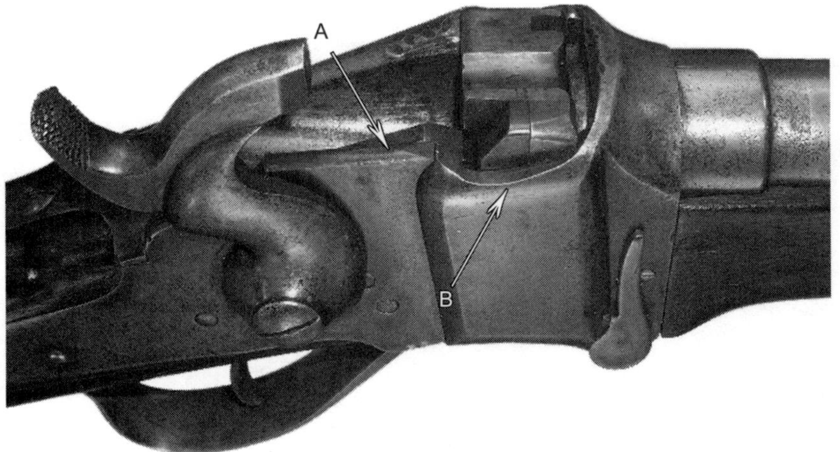

Fig. 17-12. The Type II action differed from the Type I in that Lawrence primer system was reduced to a ridge of metal (arrow A), the area around the firing pin (arrow B) was milled lower, and the new lock plate was thinner (0.19 inch as opposed to 0.38 inch).

make new barrels, this issue was resolved by fabricating a simple stepped collar 1.24 inches in diameter that threaded into the receiver. The U.S. Model 1868 barrel, which had a diameter of 1.092 inches, was then welded into the collar. The welded joint is marked by a visible change in the grain structure of the steel, causing a shadowy ring on the barrel about 0.94 inch ahead of the receiver joint, see Figure 17-13, arrow.

Fig. 17-13. The Springfield barrel blanks were too small for the Sharps receivers, so a welded "adapter" was used. The joint line (arrow) is visible for approximately 0.94 inch ahead of the shoulder.

The barrel length was 35-1/16 inches for both the Type I and II rifles and tapered from 1.092 to 0.775 inch, breech to muzzle. They were rifled with three broad grooves making one turn to the right in

42 inches. This produced an arm of the same basic overall length as a U.S. Model 1868 or U.S. Model 1870 Springfield Rifle based on the Allin pattern. Two barrel bands held the forend to the barrel, rather than three as in the previous percussion Sharps rifles.

The 22-inch-long Type I carbine barrel was held by a single band and tapered to 0.792 inch at the muzzle. It had a fixed front sight base with removable steel blade of the type used on the U.S. Model 1870 Springfield, the U.S. Model 1871 Ward-Burton, and U.S. Model 1870 Remington carbines.

The barrels were stamped with the receiver serial number, above the wood on the left side, ahead of the receiver. As the Type I rifles and carbines used percussion Sharps model receivers, they were already serial numbered. That same number was stamped on the left side of the barrel, just ahead of the receiver face, see Figure 17-14. The Type II rifles used newly manufactured actions which were serial numbered consecutively from 1 to 300. The barrels were also stamped with the receiver serial number minus the "C" prefix, see Figure 17-15.

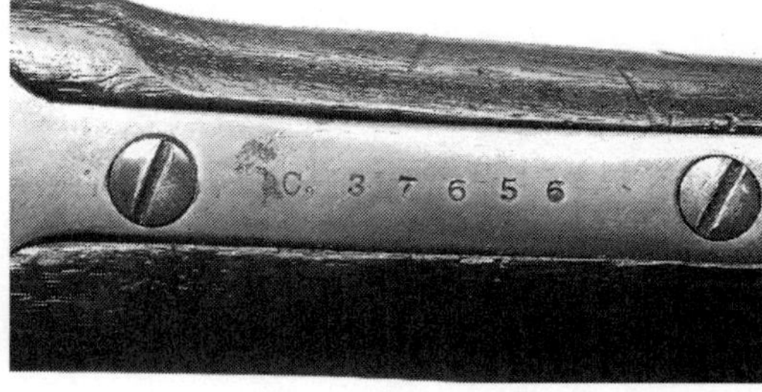

Fig. 17-14. The Springfield Armory stamped the numbers only of the old Sharps serial number on the left side of the barrel and the upper tang. Note pin in the rear sight leaf which prevented it from being deformed when folded down.

Rear Sight

Both Sharps models used the same modified Model 1868/70 rear sight with the pin added to the leaf that was also installed on both the Remington and Spencer rifles, see Figure 17-16.

Furniture

The U.S. Model 1863 rifle musket butt plate, marked "U.S." on the tang, was used on the U.S. Model 1870 Type I rifles and carbines and the Type II rifles.

The barrel bands and nose cap were essentially identical to those used on the U.S. Model 1870 Springfield Rifle.

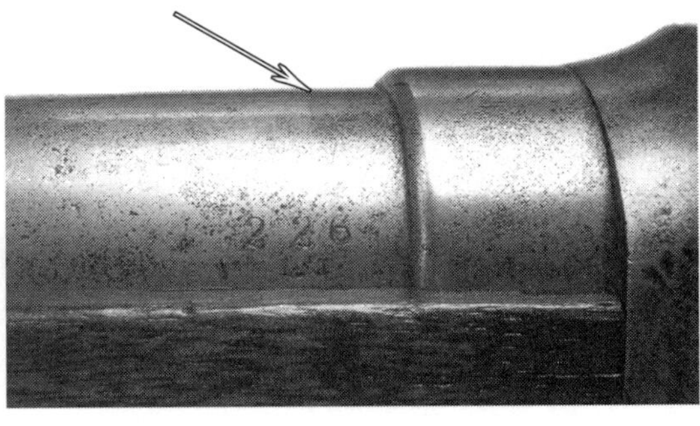

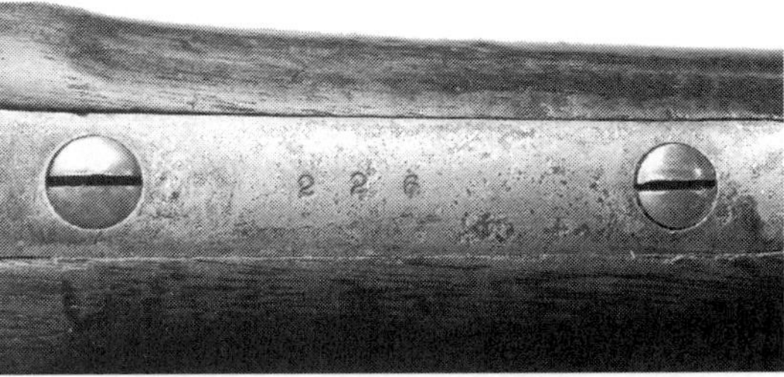

Fig. 17-15. The Springfield Armory stamped a new serial number on the left side of the Type II barrel to match that on the upper tang. The serial numbers ran from 1 to 300, the total of the U.S. Model 1870 Type II Sharps Rifles made. Note that the welded barrel adapter joint (arrow) is nearly invisible on this rifle.

Fig. 17-16. The U.S. Model 1868 rear sight was fitted to both the Type I and Type II rifles.

The bands were marked "U" on the right side. The upper band carried the forward sling swivel. The barrel bands were retained by barrel band springs. The pin of the upper barrel band spring also secured the ramrod stop for the ramrod.

The lower sling swivel was inletted into the buttstock, as the design of the receiver precluded its placement of the front of the trigger guard. It is the same rear sling swivel as used on the U.S. Model 1871 Springfield-Spencer rifle, see Figure 17-17.

Fig. 17-17. The lower sling swivel had to be mortised into the buttstock. It could not be mounted on the bottom of the action because that space was occupied by the finger lever.

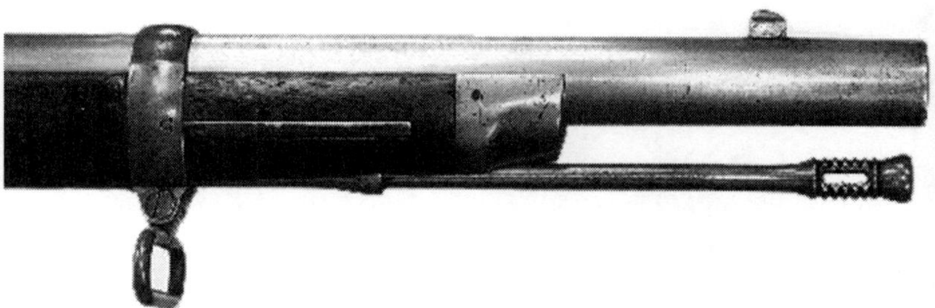

Fig. 17-18. The single-shoulder Model 1868–style ramrod was used on both the Type I and Type II rifles. When stowed, it was nominally flush with the muzzle.

Ramrod

The ramrod was added to the rifle at time of conversion. It was the U.S. Model 1868 pattern with the single shoulder and the 7-ringed head. The shank was plain without threads or cannelures. The rod was 32-3/8 inches long and extended to within 1/16 inch of the muzzle (nominally flush) when stowed, see Figure 17-18. The same ramrod was used for both the Type I and Type II rifles. The Type I carbine was not equipped with a ramrod.

Chapter 18: U.S. Model Ward-Burton Rifle Musket, Experimental Model, 1871 (Ward-Burton Patents)

Historical Background

This early bolt-action rifle design, chambered for the U.S. Government .50-70 cartridge, was the invention of Bethel Burton; General W. G. Ward was his financial backer, see Figure 18-1. Unfortunately, the Ward-Burton design was somewhat fragile; this, coupled with an unconventional appearance, meant that it did not fare well in the reports of the officers whose troops used them. The Ward-Burton rifles were also considered dangerous in the hands of untrained troops because, unlike the "trapdoor," the simple act of loading left them ready to fire, and the method of applying the safety was awkward.

Even so, the Ward-Burton design did see field service with the U.S. Army. Ward-Burton carbines, along with Model 1870 Springfield carbines (Chapter 9), are known to have accompanied Custer's 7th Cavalry on the Yellowstone Expedition of 1873, see Figure 18-2. A few were also used, with very poor results, during the Modoc War, in Northern California in 1873. Some of the Ward-Burton carbines are also said to have been issued to the army's Indian scouts after being withdrawn from trial. These are usually found in quite poor condition (see the note below regarding stock mutilation).

A handful of Ward-Burton rifles redesigned to include a tubular magazine beneath the barrel and chambered

Fig. 18-1. The U.S. Model 1871 Ward-Burton Trials Rifle, a single-shot rifle in .50-70 caliber. One of approximately 1,000 made.

Fig. 18-2. The U.S. Model 1871 Ward-Burton Carbine, a single-shot carbine in .50-55 caliber. One of approximately 300 made.

for the U.S. .45-70 cartridge were submitted for the 1878 magazine rifle trials. But they were withdrawn after severely jamming on two consecutive days, while using government-manufactured parts. The revised rifles themselves were fabricated elsewhere.

Quantity Produced

A total of 1,013 U.S. Model 1871 Ward-Burton rifles and 317 carbines were manufactured at the Springfield Armory for use during the field trials that took place during 1871 to 1872. The thirteen extra rifles and carbines were probably the result of production overruns. The Ward-Burton rifles and carbines were not serially numbered.

Adapted From

The Ward-Burton action employed a rotating bolt and was a new and innovative action and stock design. But, as it was manufactured at Springfield, many standard parts such as butt plates, butt plate screws, guard bows, swivels, front and rear sights, barrel bands, band springs, ramrod, nose cap, etc., were utilized in an as-is state or only slightly modified, from the vast supply of U.S. Model 1863 rifle musket and/or U.S. Model 1868 and 1870 components available.

Table 18 U.S. Model Ward-Burton Rifle Musket, Experimental Model, 1871	
Finishes	
Receiver	Blackened*
Breechblock or Bolt	National Armory Bright
Barrel	Rifle and Carbine: National Armory Bright

The .58- and .50-Caliber Rifles

Furniture	All National Armory Bright unless otherwise noted Rear sight: Browned (blued)
Stock	Oil-finished American Black Walnut
Markings	
Receiver	Eagle clutching arrows "U.S./SPRINGFIELD 1871" on left side, above wood line
Breechblock or Bolt	Ward Burton patent dates on top surface of bolt rib
Barrel (all)	None (on production models)
Furniture	Butt Plate: "U.S." Bands (2 for rifles, 1 for carbine): "U" near upper edge, right side Rear Sight (rifle): "2," "3," "5," "7," "9" right rear face of leaf Rear Sight (carbine): "2," "3," "5," "7" only; on leaf
Stock	"ESA" in oval on left wrist. No date. Three varying initials in banner on left side, below receiver marking. "BL" in box behind trigger guard plate
Principal Dimensions	
Overall Length	Rifle: 51-7/8 inches Carbine: 41-5/16 inches
Stock Length	Rifle: 48-3/4 inches. Carbine: 29-3/4 inches
Barrel Length	Rifle: 32-5/8 inches; 3 grooves, 1 turn in 42 inches to the right Carbine: 22 inches; 3 grooves, 1 turn in 42 inches to the right
Muzzle Diameter	Rifle: 0.775 inch Carbine: 0.785 inch
Ramrod Length	Rifle only: 35-5/8 inches (flush with the muzzle when stowed)

* The term "blackened" is used to describe the color achieved by case-hardening in oil and water. The oil was floated on water. When the part to be case-hardened was removed from the oven, it was immediately immersed in the oil-water bath. The oil produced the black color and the water hardened the steel.

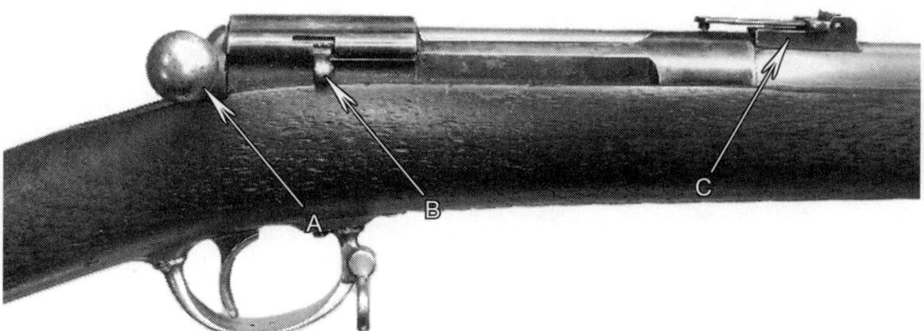

Fig. 18-3. Note the massive bolt handle (arrow A), safety lock (arrow B), and the U.S. Model 1868 rear sight (arrow C).

Identifying Features

Stock

The full-length stock was made of American black walnut, as were all stocks manufactured at the Springfield Armory in the 19th century. The rifle stock was 48-3/4 inches in length and the carbine stock was 29-3/4 inches long. The carbine stock followed the general Springfield profile regarding the shape of butt, comb, and wrist. Since the action used a rotating bolt with internal striker, no lock plate was needed, see Figure 18-3. The large-diameter, round bedding groove for the action extended straight out through the rear of the stock, see Figure 18-4. This substantially weakened the wrist but the large clearance trough was necessary, as the rounded rear surface of the bolt was the same diameter as the action and was located at the end of the receiver. Two barrel bands were used to secure the forend to the rifle barrel but only one was used on the carbine forearm.

The carbine stock was mortised for a sling bar, see Figure 18-5. It differed from the sling bar on the U.S. Model 1870 (trapdoor) carbine in that the long axes of the oval bases formed a straight line with the bar. Since there was no lock plate, the bar was mounted with two small wood screws, which were interchangeable with the rear trigger guard screw.

Fig. 18-4. The heavy bolt handle is closed and locked. Notice the large clearance groove that was required in the top of the wrist which weakened the stock but was necessary for operation.

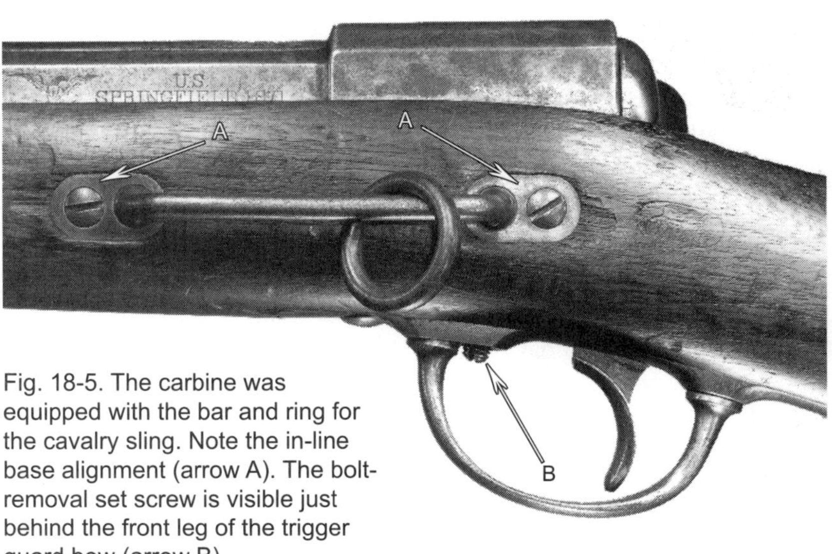

Fig. 18-5. The carbine was equipped with the bar and ring for the cavalry sling. Note the in-line base alignment (arrow A). The bolt-removal set screw is visible just behind the front leg of the trigger guard bow (arrow B).

The Ward-Burton rifle and carbine were inspected by Master Armorer Erskine S. Allin at the Springfield Armory; his "ESA" cartouche, in small oval, was struck into the left wrist. Ward-Burton stocks were also marked with an additional, banner-shaped cartouche containing various inspector's initials just below the top edge of the stock on the left side: "AGB" for A.G. Bennett; "GH" for George Hosmer (a distant cousin of the author); either "JWK" for John W. Keene or "ECW" for E.C.Wheeler.

The forearm of the carbine stock did not have a nose cap. The nose of the forearm was 3-1/2 inches long ahead of the barrel band cut. The U.S. Model 1863 nose cap was used on the rifle stock which was also cut for the U.S. Model 1868 ramrod with the single shoulder and ramrod stop which was held in place by the upper band spring pin.

NOTE: The author has observed an odd mutilation in two of the five Ward-Burton carbines examined: a set of crudely hand-gouged "grasping grooves" in the forearm. These were definitely added after manufacture, and probably after the carbine left military service.

Action

The Ward-Burton action was operated by a rotating bolt locked by the interrupted screw principle. It was single shot and the bolt handle had to be turned up 90 degrees and drawn back for a cartridge to be inserted into the chamber, refer to Figure 18-3. The bolt handle was massive and teardrop shaped and located at the rear of the action. There were visual similarities between this arm and the contemporary Brown-Merrill rifle, which the inventors would probably have seen, but the mechanism worked differently in the two designs.

On the left side of the receiver above the line of the stock was stamped a spread-winged eagle clutching arrows. Immediately to the left was a two-line marking, see Figure 18-6:

U.S.
SPRINGFIELD 1871

Fig. 18-6. The left-side receiver markings are nearly covered by the line of the stock. Note the second banner-style cartouche of the Ordnance inspector.

On the upper right rear of the Ward-Burton receiver was a small spring-loaded plunger. It had a knob, with a checkered rear face, and was mounted parallel to the bore in a slotted cylindrical recess closed at the front with a set screw, see Figure 18-7, arrow. When the bolt was raised approximately 35 degrees from its (horizontal) closed position, the plunger engaged a hole drilled in the front of the bolt handle,

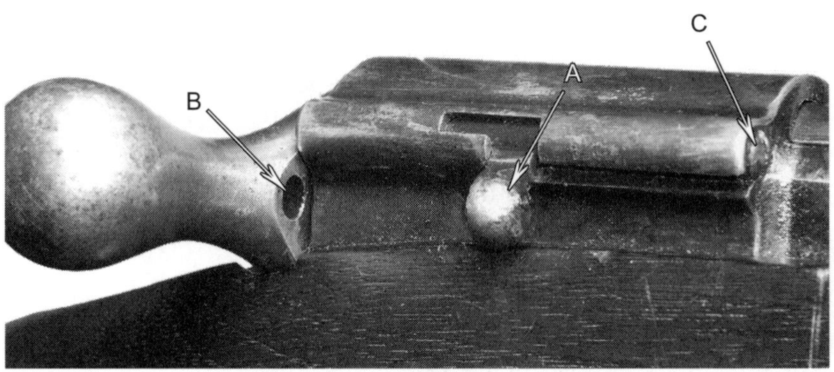

Fig. 18-7. This close-up of the safety/bolt lock (arrow A) shows that when the bolt handle is raised slightly, a spring-loaded pin can enter the hole visible at the root (arrow B). Effective but very fragile, the small checkered knob at the rear is often broken off. Arrow C shows the set screw which retains the safety lock spring.

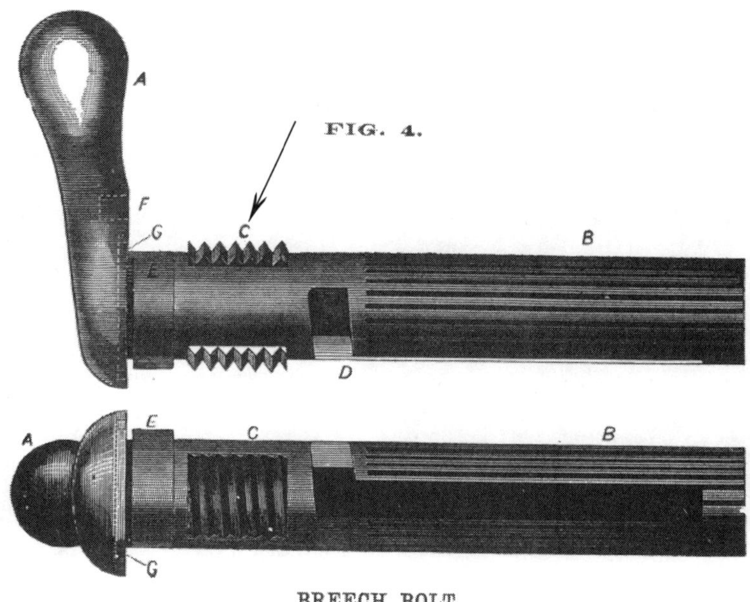

Fig. 18-8. The original manual for the Ward-Burton rifle, published in 1872, contains this illustration of the bolt body. Note the interrupted locking threads at "C" (arrow).

near the body, thus functioning as a safety, and/or a bolt lock. This part was quite fragile and broke easily at the knob. It is often missing entirely on examples found today.

The action was cocked on the closing stroke of the bolt. The bolt was locked in battery in the receiver by two sets of six interrupted threads located in the rear of the receiver, just ahead of the root of the bolt handle, see Figure 18-8. The bolt body had twenty-one shallow longitudinal flutes, 2-3/4 inches long, presumably to reduce friction, and/or retain lubricant.

The loading/ejection port in the top of the receiver was filled by a large rib when the bolt was closed. This rib was dovetail-shaped in cross section, with a rounded top and front end. It carried the two-line inscription, reading from muzzle to breech, see Figure 18-9:

WARD.BURTON.PATENT.
DEC.20.1859.FEB.21.1871

Fig. 18-9. The Ward-Burton patent markings are stamped on the top rib of the bolt.

Among the distinctly "modern" features of this arm were the recessed bolt head which completely enclosed the rim of the cartridge case, see Figure 18-10, arrow A; a bolt-mounted hook extractor (arrow B); and what is believed to be the first-ever use of a "floating"

Fig. 18-10. This view of the bolt face shows the full recess for the head of the cartridge (arrow A), the hook extractor (arrow B), and the floating ejector pin (arrow C). The set screw which holds the safety plunger is also visible (arrow D).

plunger-type ejector, set into the bolt face (arrow C). Other modern features were the positive firing pin retraction and constant sear engagement, regardless of bolt handle position, from partial closure until discharge.

Some unique features are also found on the trigger guard plate. The forward trigger guard screw was a large oval-head machine screw which threaded into a lug on the bottom of the receiver. This replaced the wood screw used on most rifles of the period and firmly seated the receiver in the stock, see Figure 18-11, arrow A.

The upper front portion of the trigger formed a thin, flat bar which was attached with a small screw to the vertically sliding sear. A long headless set screw was installed inside the guard bow, just in front of the trigger. This screw, refer to Figure 18-11, arrow B, which extended approximately 3/16 inch (0.19 inch) above the guard plate,

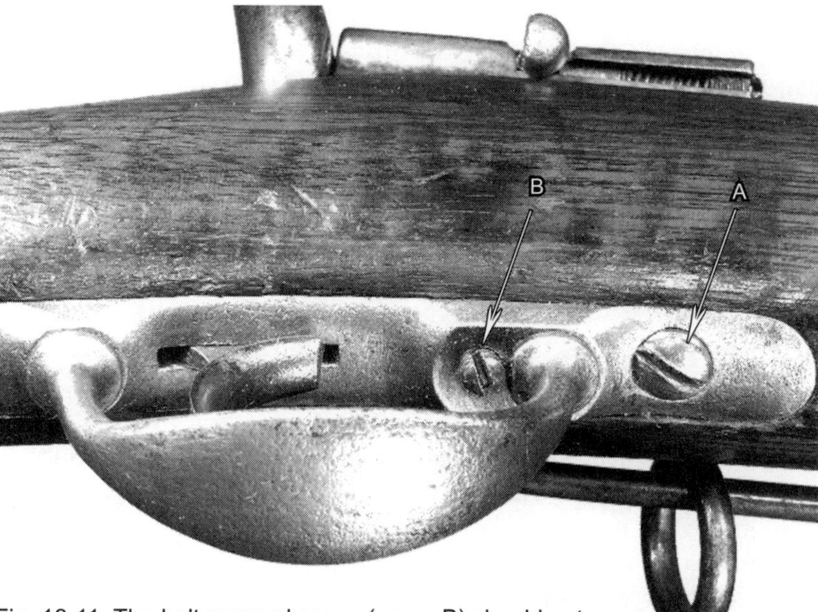

Fig. 18-11. The bolt removal screw (arrow B) should not be confused with the forward trigger guard screw (arrow A) that secures the trigger assembly to the stock. Note also the narrow trigger guard bow which is found only on the carbine.

had a part of its threaded shank cut away at the top. When the slot in the head of the set screw was at right angles, or parallel, to the bore, the trigger could not be pulled far enough to the rear to release the firing pin.

With the set-screw slot properly oriented for firing, the trigger can be pulled back allowing its front "bar" portion to drop into the area where metal was removed from the screw. This action withdraws the sear below the boltway, and permits the arm to be fired or the bolt to be pulled out. You may have to find the "correct" slot orientation by trial and error.

On the author's two Ward-Burtons, this happens to be the "10 to 4 o'clock" position, when the muzzle is pointing up, and two threads are exposed. Of course, removing the trigger guard will quickly reveal exactly how the above described parts interact with each other.

Once removed, the bolt is easily disassembled, without any tools, simply by rotating the handle 90 degrees to the right, when viewed from the rear. The mainspring is not powerful at all, but attention should be paid to the manner in which the various parts (body, sleeve, firing pin, mainspring, extractor, and ejector) all nest together.

Barrel

The rifle barrel was 32-5/8 inches long; the carbine barrel was 22 inches long. The U.S. Model 1871 Ward-Burton was chambered for the .50-70 caliber, centerfire metallic cartridge.

Barrels were rifled with three broad lands and grooves, one turn to the right in 42 inches. The U.S. Model 1868 rear sight was mounted 1/32 inch (0.03 inch) ahead of the receiver. The only markings on the barrel are a set of witness marks visible in the scallop behind the rear sight. Most were finished bright, but a few were blued.

A unique U.S. Model 1871 Ward-Burton carbine is in the collection of the Springfield Armory Museum. It bears the following marking on the top of the barrel, ahead of the rear sight:

Eagle head
Crossed arrows
WARD BURTON CARBINE
BARREL
1871

These marks *do not* appear on production carbines.

Furniture

The U.S. Model 1863 rifle musket butt plate was used on the both the U.S. Model 1871 Ward-Burton rifle and carbine. The rifle forend used the U.S. Model 1863 nose cap. The forend was secured to the barrel by two Model 1870 barrel bands and band springs. The carbine forearm was secured by a single barrel band and band spring.

Rear Sight

The U.S. Model 1868 rifle rear sight was installed on the U.S. Model 1871 Ward-Burton rifle. It was marked with gradations for "2," "3," "5," "7," "9" hundred yards. The elevation gradations were marked on the right side of the leaf. The rear sight used on the carbine was identical to that used on the U.S. Model 1870 Springfield carbine (described in Chapter 9). It was marked only for "2," "3," "5," and "7" hundred yards. The leaf slide was 0.169 inch in height. A pin, threaded into the top rear face of the sight leaf, protected it from being bent while folded down.

Front Sight

The Ward-Burton rifle was equipped with the U.S. Model 1868 front sight and base machined as one unit and which also acted as a bayonet stud, see Figure 18-12. This front sight design was common to the U.S. Model 1868/70 .50-70 Springfield arms, as well as the balance of the trial rifles described in previous chapters.

The carbine used a two-piece sight (base and blade) to allow for blade replacement. The front sight blade was subject to wear when being drawn in and out of the carbine sockets, and by the action of dust and

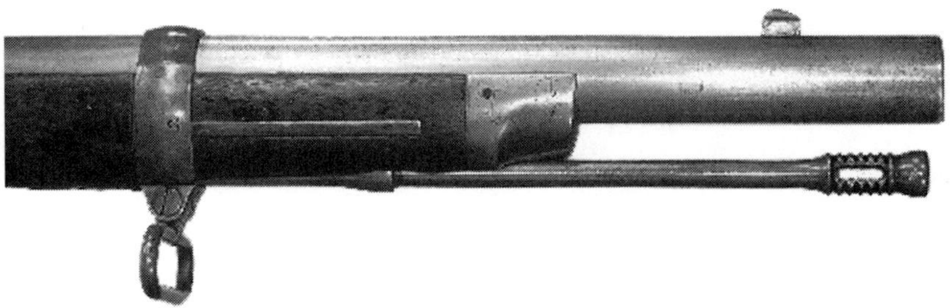

Fig. 18-12. The Ward-Burton rifle used the Model 1870 upper band with swivel, a Model 1863 nose cap and the Model 1868 ramrod with the single shoulder. The front sight was made in one piece and also served as the bayonet stud.

dirt kicked up by horses. Unlike the U.S. Model 1870 Springfield carbine front sight, which had a brass blade, the Ward-Burton front sight blade was steel, see Figure 18-13 arrows. All subsequent front sight blades made at Springfield would also be made of steel.

Ramrod

Fig. 18-13. The carbine front sight was made in two pieces, a steel stud (arrow A) and a separate steel blade (arrow B) that could be replaced when worn.

The ramrod used with the U.S. Model 1871 Ward-Burton rifle was the U.S. Model 1868 ramrod with seven rings around the head and a single locking shoulder. It was finished in the white and the shank was plain without threads or cannelures. Its length was 35-5/8 inches and it was 1/16 inch short of the muzzle (or nominally flush) when stowed, refer to Figure 18-12.

The ramrod was not used with the carbine.

Cartridge Carriers/Boxes

Rifle Cartridge Carriers

The ammunition carriers issued with the U.S. Model 1865 and 1866 Rifles were Civil War–era cartridges boxes; for rifles, they were generally the Pattern 1864 cartridge box, but earlier versions are known to have been used as well. A number of conversion methods were tried to develop a cartridge box that would be both practical and efficient in service.

Five basic types of cartridge boxes were issued for use with the rifle musket between 1861 and 1865. The first two were developed for carrying .69-caliber paper cartridges; the latter two were for carrying .58-caliber paper cartridges. The cartridges were held in two tin inserts. The exact pattern can be determined by the measurements shown in Table 19-1.

Table 19-1
Civil War Rifle Musket Cartridge Boxes
(Inches)

Pattern	Length	Width	Depth	U.S. Insignia/ Other
1839 (.69)	7.2	1.6	5.8	Brass Plate
1857 (.69)	8.2	1.8	4.8	Brass Plate
1857 (.58) Type I	7.2	1.7	4.3	Brass Plate
1857 (.58) Type II	7.2	1.7	5.1	Brass Plate
1861 (.69) Type I	7.8	1.6	4.7	Brass Plate
1861 (.58) Type II	6.8	1.4	5.2	Brass Plate
1864 Type I (.58)	7.2	1.7	4.3	Brass Plate/small buckles sewn with curved stitching
1864 Type II (.58)	7.2	1.7	4.3	Embossed U.S. on flap

The .58- and .50-Caliber Rifles

The Pattern 1861 Type II (.58 caliber) or 1864 Type II (.58 caliber) leather cartridge boxes were issued with the U.S. Model 1865 rifle. They were modified for use with the .58-60 metallic self-contained cartridge by the insertion of a wooden block in the top of each tin insert. The wooden block held eight cartridges and an additional eight cartridges could be carried in the bottom pocket of each tin insert. The two tins thus held a total of thirty-two cartridges, see Figure 19-1.

It appears that the majority of rifle musket cartridge boxes issued with the new .50-caliber rifles were the Pattern 1864, Type II, see Figure 19-2. As they were manufactured from July 1864 to the end of the war, they would have been the most commonly available pattern in storage and in the best condition for active service.

Fig. 19-1. Pattern 1861 .58 Caliber Cartridge Box modified for use with the U.S. Model 1865 Rifle. Wooden blocks holding eight cartridges were inserted into the tops of the two tins. The bottom compartment of each held eight more cartridges for a total of thirty-two. North Cape Publications collection.

The three most common conversion methods used to modify the cartridge boxes for use with the metallic cartridges were as follows: 1) Discard the two tins used to hold the percussion rifle musket paper cartridges and insert two of the regulation cardboard boxes holding twenty metallic cartridges each. 2) Line the inside of the cartridge box with a wool pad. The cartridges could then be dumped into the cartridge box with the wool holding them snugly enough to prevent their rattling against each other. Unfortunately, in wet weather, the wool absorbed water which sometimes penetrated into the cartridge. The

Fig. 19-2. The Pattern 1864, Type II rifle musket cartridge box was issued to infantry troops during the Civil War to hold paper cartridges for the rifle musket. Craig Riesch collection.

wet wool also made it difficult to remove the cartridges from the cartridge box. 3) Discard the tins and insert a wooden block sized and drilled with twenty holes to hold the .50-70 caliber ammunition, see Figure 19-3.

In addition to converted Civil War rifle musket cartridge boxes, in 1866, the Ordnance Department conducted experiments with a number of newly made cartridge box designs. One contained a wooden block drilled to hold twenty .50-70 cartridges. The inside dimensions of this cartridge box were 6 inches long by 1.5 inches wide by 3.5 inches high. It was made of black bridle leather and had two vertical and two horizontal loops on the back for both the shoulder sling and the waist belt. The box examined was marked:

U.S. ARSENAL
SPRINGFIELD
MASS.
1867

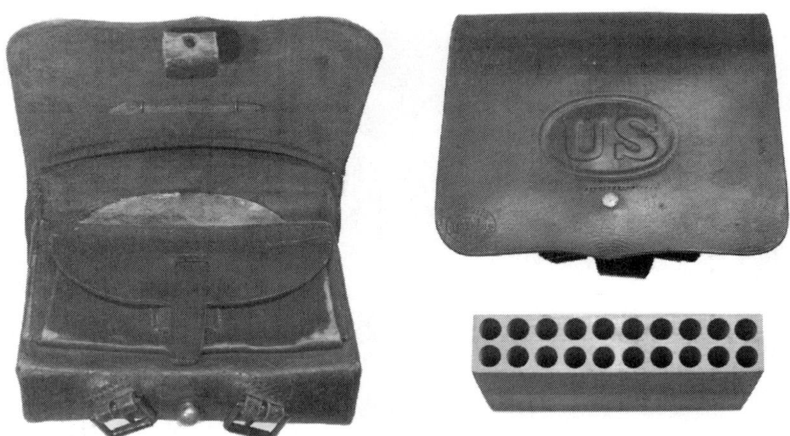

Fig. 19-3. Pattern 1861 and 1864 cartridge boxes were modified for the .50-70 cartridge by (left) the installation of a woolen lining (Bill Chachula collection) or (right) the substitution of a wooden block for the paper-cartridge tin holders that held twenty cartridges (Craig Riesch collection).

Another cartridge box was designed and made at Watervliet Arsenal. It was made of black bridle leather with a single flap; it was 7.1 inches long by 1.6 inches wide by 2 inches high and held a single tin insert with holes top and bottom for nineteen .50-70 cartridges. The cartridge box was worn on the belt, suspended from two vertical loops sewn to the back of the box. The flap was stamped:

US
WATERVLIET
ARSENAL

A third experimental cartridge box employed the Pattern 1864, Type II cartridge box to hold a block of wood drilled for twenty cartridges in each end. The block was mounted on a slide and swivel. When the top twenty cartridges were used up, the block was pulled along its slide and turned upside down on its swivel, then pushed back down in the cartridge box to present twenty more cartridges. This might have been a practical alternative but was too costly to produce.

It is not known if the Blakeslee Quick Loader was intended to have

Fig. 19-4. The Blakeslee cartridge box for the Spencer carbine and rifle was issued in the post–Civil War period for use by Cavalry troopers armed with the Spencer carbine. As the Model 1871 Spencer rifle was not issued, it is doubtful if the Blakeslee cartridge box was still in use after the Spencer carbines were withdrawn from the regulars by early 1873. Photo courtesy of Roy Marcot.

been issued with the Pattern 1871 Spencer rifle. The reports entitled *Summary Statement of Ordnance and Ordnance Stores on Hand in the Cavalry Regiments of the United States* for the period 1867 through 1872 show that the Spencer carbine in both .52 and .50 caliber was in use, and notations on surviving copies show that the "Blakeslee" cartridge box had been issued to at least a few companies. It is therefore possible it would have been intended to be used with the U.S. Pattern 1871 Springfield-Spencer Rifles if they had ever been issued.

The Blakeslee Patent Spencer carbine box was 12 inches high and made of black bridle leather. It held either six, ten or thirteen tubes, each of which contained seven cartridges. The six-tube patterns were rectangular, and the ten- and thirteen-tube patterns were hexagonal, in cross section. The six- and thirteen-tube patterns had a separate tool pouch sewn to the front but the ten-tube pattern did not, see Figure 19-4. All three patterns were worn on shoulder straps. Only 500 six-tube and 1,000 thirteen-tube patterns were manufactured; at least 32,000 of the ten-tube patterns were manufactured and delivered to the Ordnance Department by June 30, 1866.

As the Spencer carbine was withdrawn from use by mounted troops, the utility of the Blakeslee cartridge box declined. The U.S. Government held an auction on July 3, 1869, for various military stores, included in which were 5,000 cavalry-style (ten-tube) Blakeslee cartridge boxes. Roy Marcot (*Spencer Repeating Fire*arms) states that "the last Blakeslee box must have been sold to surplus houses by 1877 . . ."

Fig. 19-5. This Civil War cartridge box issued with the Spencer, and later, the Joslyn carbines, continued in use in the post-War period until the Spencer carbines were withdrawn. They may also have been intended for issue with the U.S. Model 1871 Infantry (Spencer) Rifle. Craig Riesch collection.

A more likely candidate for issue with the post–Civil War Spencer carbines, and the U.S. Pattern 1871 Spencer Rifle was the Spencer Rifle or Carbine Box, see Figure 19-5, that was also issued for use with the Joslyn carbine. It was made of black bridle leather and the interior was 7.1 inches long by 1.5 inches thick and 2.4 inches high and held six packets of seven cartridges. Another variation is known that measured 8.5 inches wide by 1.5 inches thick and 2 inches high. These appear to have been under federal government contract during the Civil War by E. Gaylord, J.T. Pittman, H.A. Dingee and others.

The "Hagner No. 1" Cartridge Box

This cartridge box, and the "No. 2" box described below, were designed by Colonel Paul V. Hagner of the Ordnance Department in 1872. They were intended to be worn in pairs, usually two of the same pattern. But it is also definitely known that one of each pattern was occasionally issued at the same time. The larger No. 1 size was originally intended for infantry use, the smaller No. 2 for mounted use, but both sizes saw infantry use.

Fig. 19-6. The Hagner No. 1 cartridge box was designed by Col. Paul V. Hagner of the Ordnance Department.

The "No. 1" box was made of dark brown leather, 6-3/8 inches wide by 1-1/4 inches high and 1-5/8 inches deep. It had a tool pocket that protruded 1/2 inch on the left side. The box was closed by a single cover flap scalloped at the bottom in the form of a shield, 6-7/8 inches wide, see Figure 19-6. The cover flap was a continuation of the rear wall of the box. It folded over the top of the box and fastened at the bottom front with a stitched and riveted tab, 7/8 inch wide by 4 inches long. The tab engaged a brass finial on the bottom edge of the box. The cover flap was embossed "US" in an oval, 2-3/4 inches wide by 1-9/16 inches high. The end of the tab was marked with inspector's initials, such as "T.F.F."

Fig. 19-7. This rear view of the Hagner No. 1 box shows its "Y"-shaped strap through which the leather garrison belt was slipped.

The suspension at the rear of the Hagner No. 1 cartridge box was a "Y"-shaped leather yoke attached at the upper rear of the box with two rivets on each side, see Figure 19-7. The yoke folded over on itself and was restrained by a horizontal strap riveted at either end. The restraint strap was 4-5/8 inches long by 5/8 inch (0.63 inch) wide. When worn on the waist belt, the yoke was routed under the restraint strap and attached to the finial on the bottom of the cartridge box that also fastened the flap closed. The waist belt was passed through the top of the "Y." The box was marked "No 1" at top center, rear, see Figure 19-8.

Fig. 19-8. The Hagner cartridge boxes were stamped on the back "No 1" or "No 2" at the top, center, depending on the pattern. These marks are often worn away and cannot be read. See also Fig. 19-15.

When the cover flap was raised, the cartridge box was tipped upward about 2-1/8 inches to make it easier for the soldier to remove a cartridge. Three rows of eight loops held a total of twenty-four cartridges. One row of eight loops was stitched to the top, rear wall of the cartridge box. Two rows of eight cartridge loops each were stitched to the rear wall at the bottom, see Figure 19-9. The loops were 1 inch high and stitched to a web backing 1-1/2 inches high.

Fig. 19-9. The Hagner No. 1 box held twenty-four cartridges in three rows of eight. Two rows were stitched to the bottom walls.

Supplies of the Hagner No. 1 cartridge box continued in use for some years after the adoption of the .45-70 cartridge, but were mostly issued to support units or civilian contractors. The .50-70 boxes were also modified at Rock Island Arsenal to hold .45-70 ammunition. A small strip of leather was sewn across the width of each row of cartridges. The strip was sliced so as to form "fingers" which were pushed down into the loops to make them tight enough to hold the smaller-diameter .45-70 cartridge.

Carbine Ammunition Carriers

Ammunition carriers issued with the trial carbines appear to have either been the Pattern 1855 Carbine Cartridge box, or a box based on that pattern, refer to Figure 19-5. The tin inserts originally intended to hold paper cartridges were discarded and replaced with either a wool lining or a single pine block bored for twenty cartridges.

Cartridge boxes made for the Sharps percussion pattern carbines during the Civil War were also issued with the various carbines in use between 1865 and 1872 according to availability, see Figure 19-10. They were modified with wool inserts or wooden inserts to hold the cartridges, see Figure 19-11. These boxes, in original Civil War configuration

Fig. 19-10. The Pattern 1855 carbine cartridge box was issued for use by mounted troops armed with carbines. It continued to be issued until the advent of the Dyer pouch in 1872. Two variations are shown here: above, with the original buckles retained; below, with the unnecessary buckles removed. Craig Riesch collection.

Fig. 19-11. The tins used to hold the paper cartridges were removed from the Pattern 1855 cartridge box and replaced with a wooden block drilled to hold twenty .50-70 cartridges.

with a single tin tray, were also issued occasionally. Reports indicate that the troopers simply discarded the tin trays, replacing them with cardboard boxes of cartridges.

The Blakeslee Quick Loader as well as the Pattern 1855 Carbine Cartridge boxes were used with the Pattern 1865 Spencer carbines that were issued to mounted troops, teamsters and others until replaced by the Pattern 1873 and Pattern 1877 .45-70 Springfield carbines.

Dyer Cartridge Pouches

From 1870 to 1874, several versions of this pouch, designed by Brigadier General Alexander B. Dyer of the Ordnance Department, were produced. The very first issue combined a carbine sling with the cartridge pouch, see Figure 19-12. Only 100 were issued for field trial.

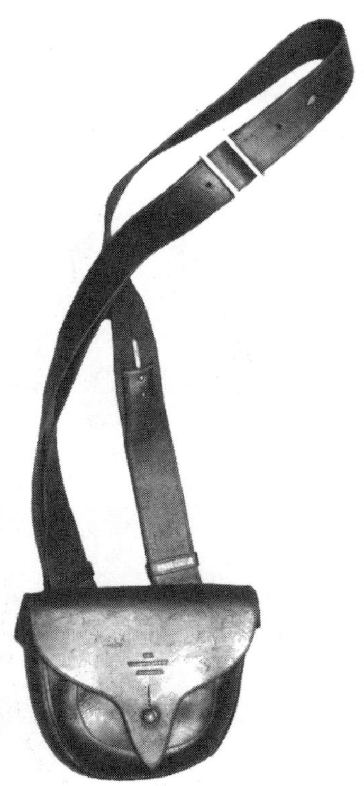

Fig. 19-12. One hundred of the experimental Pattern 1870 Dyer Cartridge Pouches with the shoulder sling for the carbine were manufactured and sent to the 1st Cavalry for trial. Photo courtesy of R. Stephen Dorsey.

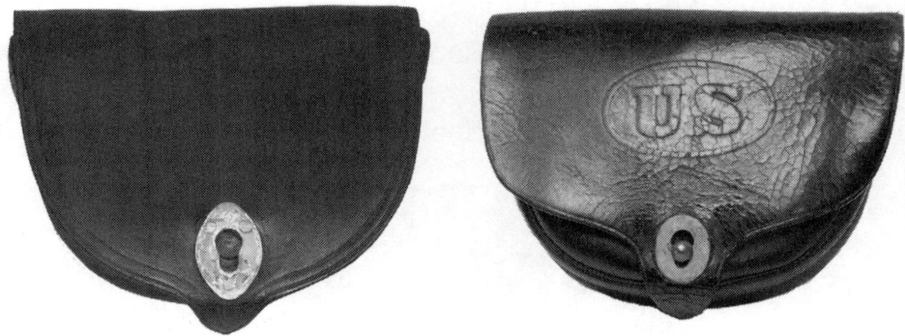

Fig. 19-13. Left, the Pattern 1872 Dyer Pouch flap was unmarked. Right, the Pattern 1874, as well as later-pattern Dyer Pouches, was stamped "U.S." in an oval on the flap. The Pattern 1872 was designed to hold the .50-70 cartridge, the Pattern 1874 to hold both the .50-70 and .45-70 cartridges. The Dyer Pouch was issued to mounted troops. Left, Craig Riesch collection; right, author's collection.

This pattern could also be fastened directly to the waist belt.

The issue cartridge pouch was the Pattern 1872 Dyer Pouch. It had a large rear compartment lined with sheepskin to carry up to sixty .50-70 cartridges for the U.S. Model 1867/68 Sharps and the U.S. Model 1870 Springfield carbines. The pouch was shaped like a half moon with a flap sewn to the back and folded over the top. It was closed with a finial and escutcheon at the bottom front, see Figure 19-13.

The large rear pocket was fleece-lined. The smaller, unlined front compartment held the issue combination tool or tools or several pistol cartridges. The marking, of the U.S. Arsenal at Watervliet, N.Y., was stamped inside the pouch:

U.S.
WATERVLIET
ARSENAL

The Pattern 1874 Dyer Pouch was marked "U.S." in large letters stamped within an oval on the flap. The Pattern 1872 can be identified by the long, sharply pointed flap and Watervliet Arsenal marking on the inside. It was also issued for use with the .50-70 cartridge.

The "Hagner No. 2" Cartridge Box

The "Hagner No. 2" box was also made of dark brown leather, see Figure 19-14. It was 6-1/8 inches wide, 3-1/4 inches high, and 1-7/8 inches deep. It also had a small tool pocket that protruded 1/2 inch

Fig. 19-14. The Hagner No. 2 cartridge box was developed for mounted use, but in fact was worn by infantry as well since it was compact and handier for belt use. It had two simple belt loops on the back.

on the right side. Like the Hagner No. 1 Cartridge Box, the cover flap was integral with the back wall and was scalloped at the bottom in the form of a shield. It was 7-1/4 inches wide and fastened at the bottom front by means of a stitched and riveted tab, 7/8 inch by 3-3/8 inches, which engaged a brass finial, on the bottom edge of the box. The cover flap was embossed "US" in an oval, 2-3/4 inches wide by 1-9/16 inches high. The end of the tab was marked with inspectors' initials, such as "D*P."

Fig. 19-15. "No 2" was stamped on the back of the smaller Hagner cartridge box.

The rear of the box was marked "No 2," see Figure 19-15. The suspension at the rear of the No. 2 box was simpler than the No. 1 but not as adaptable. Two loops were sewn to the back of the box for the soldier's belt. The loops were 5/8 inch wide by 5 inches long and were

Fig. 19-16. The Hagner No. 2 cartridge box also held twenty-four cartridges in three rows. The third row was sewn to the outside front of the box and protected by the cover.

attached to the rear and bottom edge of the box, 3-3/4 inches apart. Each was reinforced with a rivet at top and bottom. With the lid raised, it will be noted that the front of the box comes up approximately 2-3/8 inches, see Figure 19-16. The cartridge box held twenty-four cartridges in web loops, sixteen sewn to the back of the box in two rows of eight, and a row of eight stitched to the front of the box and shielded by the cover flap.

National Guard/Militia Cartridge Ammunition Carriers

The Frazier Cartridge Box

The Frazier Cartridge Box was privately produced, see Figure 19-17.

Fig. 19-17. Frazier's Patent cartridge box was widely used by state and militia troops and is probably the most often encountered .50-caliber cartridge box of the period. This example carries a brass "NG" (National Guard) plate.

It was not issued to the regular Army but it was widely used by state militia and National Guard troops who carried the various .50-caliber Springfields. The Frazier box was issued to the New York National Guard and is probably the most frequently encountered historic .50-caliber cartridge box today.

The Frazier cartridge box was made in two separate sections. The bottom portion was fashioned from three pieces of dark brown leather. It was 7 inches wide by 1-5/8 inches deep by 2-3/16 inches high at the front, but 2-1/2 inches at the back with a flat bottom. The end pieces were therefore trapezoidal in shape. A wooden block 6-9/16 inches long by 1-1/4 inches wide, with a beveled top matching the angle of the ends, contained the cartridges in two rows of 9 holes each. The holes were bored 0.57 inch in diameter and 2 inches deep. The bevel provided sufficient offset that the top row of cartridges were easily extracted first, see Figure 19-18.

Fig. 19-18. The Frazier cartridge box held eighteen cartridges in a sloped wooden block. The markings are visible on the front of the box.

Typical markings on the front of the box were arranged in barred circle:

FRAZIER'S PATENT
FEBY 5TH 1872
REISSUED MCH 5TH 1878
PATENT APRIL 23RD 1878

Most boxes will also have a maker's stamp, usually one of the following:

McKENNEY
& CO
NEW YORK

or

RIDABOCK & CO.
SUCCESSORS TO J.H.McKENNEY & CO.
NEW - YORK

The cover or lid of the box was 7-3/8 inches wide by approximately 5 inches and likewise made of three pieces of leather. It was mounted with two 1-inch by 1-inch round-cornered brass hinges, each secured with four brass rivets, on the bottom front edge of the box. When opened, the cover fell forward out of the way, allowing the same access afforded by the later McKeever cartridge boxes. The lid was lined at the top edge with a curved strip of steel, approximately 1 inch by 7-1/8 inches and which was held in place at the center by the finial rivet, and by two half-circle leather reinforcements riveted to the ends. The lid, which overlapped the box at the ends, was closed by a leather strap, 1 inch wide by 4 inches long, stitched and riveted to the back of the box. The finial was in the shape of a flattened sphere on a stem, and the round brass escutcheon that reinforced the hole in the lid strap was 7/8 inch in diameter.

The sample Frazier cartridge box examined (refer to Figures 19-17 and 19-18) bears a generic National Guard brass plate 2-3/16 inches by 1-1/2 inches, with the letters "NG" in script stamped into it. The back of the plate was filled with lead. It was attached with two wire loops and a brass staple. Other plates with the initials of various National Guard or state militia units were also manufactured. Some examples are "NG" for National Guard, "NJ" for New Jersey, "PNG" for Pennsylvania National Guard, and "SNY" for State of New York.

NOTE: As a point of interest for collectors, lead more than 60 to 80 years old tends to oxidize on the surface and harden. If the lead shows streaks and specks of whitish oxidation and you can't scratch it with your thumbnail, it is probably an original plate. If you can scratch the lead with your thumbnail, it probably is a recent reproduction. Be sure and wash your hands carefully after handling lead.

Tools

The soldier was expected to keep his rifle or carbine clean and free of rust and corrosion and to make minor repairs in the field when necessary under the supervision of a noncommissioned officer. To this end, simple tools were provided to dismount the rifle and its lock work. In the period when the .58- and .50-caliber Springfield-made rifles and carbines were in use for both issue and trial weapons, the soldier had access to a mainspring vise as well as a band-spring punch and was issued one of two types of screwdrivers.

All tools were blued using the ash-bluing method.

Model 1855 Mainspring Vise

One Model 1855 Mainspring Vise was issued with each crate of twenty rifles during the Civil War and on into the Indian War period, see Figure 19-19. The mainspring vise was quite simple; it was composed of a fixed top jaw and a moveable lower jaw which could be turned up or down with a thumbscrew. To use, the hammer was cocked and the lock removed from the rifle. The fixed jaw at the top of the vise was placed over the short or top segment of the "V"-shaped mainspring. The thumbscrew was turned to raise the moveable jaw against the larger bottom segment of the spring until it was compressed slightly. Rocking the vise up and down loosened the mainspring pin in its seat and allowed the mainspring to be removed from the lock plate. The

Fig. 19-19. The Model 1855 Mainspring Vise was placed over the cocked mainspring and tightened to compress the mainspring far enough to lift off the lock plate.

remaining parts of the lock could then be dismounted for repair, replacement, or cleaning.

The Model 1855 Mainspring Vise was issued one per crate of twenty rifles or carbines.

Model 1863 Combination Tool

The Model 1863 Combination Tool was similar to the Model 1855 Combination Tool but lacked the third screwdriver blade, see Figures 19-20 and 19-21. The cleanout screw on the side of the bolster for the U.S. Model 1855 and Model 1861 rifle muskets had been eliminated in the Model 1863 and Model 1864 rifle muskets and so there was no longer any need for the smallest screwdriver blade. The Model 1863 Combination Tool consisted of two screwdriver blades sized for the screws in the Model 1863 and Model 1864 rifle musket and the wrench on the end of the longest blade. The open-end wrench served as a vise for the mainspring. With the lock removed from the rifle or carbine and the hammer cocked, the open end of the wrench was slipped over the compressed mainspring. The hammer was then eased forward and the mainspring removed.The Model 1863 Combination Tool was issued with the U.S.

Fig. 19-20. The Model 1863 Combination Tool was similar to the Model 1855 Combination Tool but with the third, small screwdriver blade removed.

Model 1865 .58 Caliber Springfield Rifle and the U.S. Model 1865 .50-70 Joslyn Rifle.

Fig. 19-21. The Model 1855 Combination Tool was issued with the U.S. Model 1855 Rifle Musket, Rifle, Carbine, and Pistol Carbine, and the U.S. Model 1861 Rifle Musket. Thousands were left in inventory after 1865 and were modified through the ensuing years for other rifles and carbines.

Model 1866 Combination Tool

The U.S. Model 1866 .50-70 Springfield Rifle's breechblock was attached to the receiver with a nut that required a spanner wrench with twin bits to remove. Surplus Model 1855 combination tools were modified to make this model by cutting off the wrench portion and machining two bits in to the top of the leg to form the spanner wrench. The screwdriver blades were retained, see Figure 19-22. The Model 1866 Combination Tool was issued with the U.S. Model 1866 .50-70 Caliber Springfield Rifle.

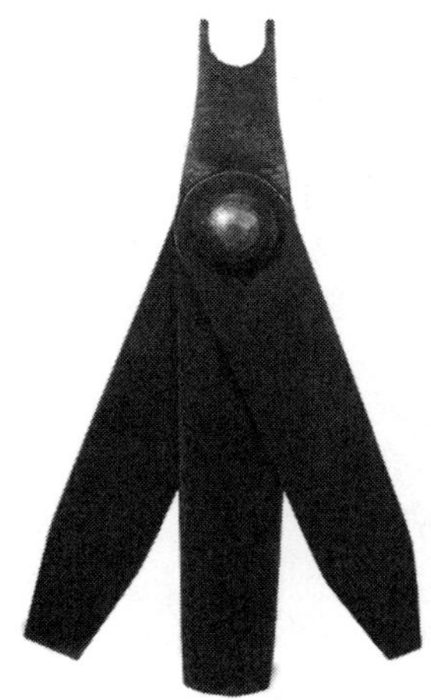

Fig. 19-22. The Model 1866 Combination Tool was manufactured at Springfield by cutting off the wrench portion of the Model 1855 tool and milling twin bits on either side to make a spanner wrench to use in removing the breechblock.

Model 1855 Band Spring and Tumbler Punch

This tool was known by several designations during its period of use, which ranged from 1855 to the end of the .45-70 Springfield era after the beginning of the 20th century, see Figure 19-23. It was used to remove the band springs from the stock, and the tumbler from the lock. The tool consisted of two small-diameter punches fastened at the top with a rivet that allowed them to swing away from each other. To dismount a band spring, the smaller-diameter punch was inserted into the small hole in the left side of the stock opposite the end of the band spring and against the band spring pin. By tapping on the end of the punch, the band spring was forced out of its groove in the stock.

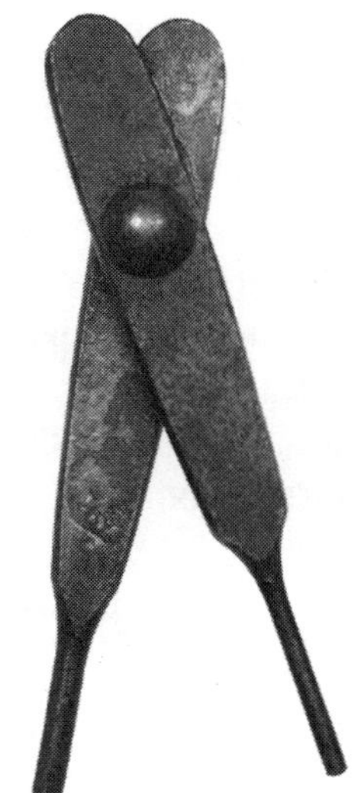

Fig. 19-23. The Model 1855 Band Spring and Tumbler Punch was used to remove the band springs and the tumbler from the hammer and lock plate.

The larger-diameter punch was used to loosen and remove the tumbler from the lock. The hammer screw was removed and the punch inserted so that it bottomed in the screw hole of the tumbler. With the lock plate lying face up across two flat surfaces so that its center area around the hammer was unsupported, the punch was tapped with a hammer to drive the tumbler shaft out of the hammer.

Four band spring and punch tools were issued with each crate of twenty rifles or carbines with all service rifles and carbines during the period, 1865–1872.

Model 1866 Band Spring and Tumbler Punch with Screwdriver

This tool was a Model 1855 Band Spring and Tumbler Punch with one of the legs modified by grinding it into a screwdriver bit at the top, see Figure 19-24. Another attempt at economizing, it was issued with the U.S. Model 1866 Rifle. Either it or the Model 1855 Band Spring Punch

Fig. 19-24. The Model 1866 Band Spring and Tumbler Punch was similar to the Model 1855 but the leg with the smaller-diameter punch was modified by grinding a screwdriver tip at the top end.

was issued with all rifles to the end of the "trapdoor" era, except for the experimental models.

Combination Screw Driver, Experimental, 1870

A great deal of confusion has arisen regarding the Combination Screw Driver, Experimental, 1870. Many collectors have noted that there are two variations, one larger than the other. It is known the larger variation was issued with the Springfield, Remington and Sharps trials rifles during the Field Trials of 1871–1872. It is believed by many collectors that the smaller variation was manufactured for issue only with the Ward-Burton rifle and carbine, see Figure 19-25. For that reason, many collectors refer to it as the "Ward-Burton Tool."

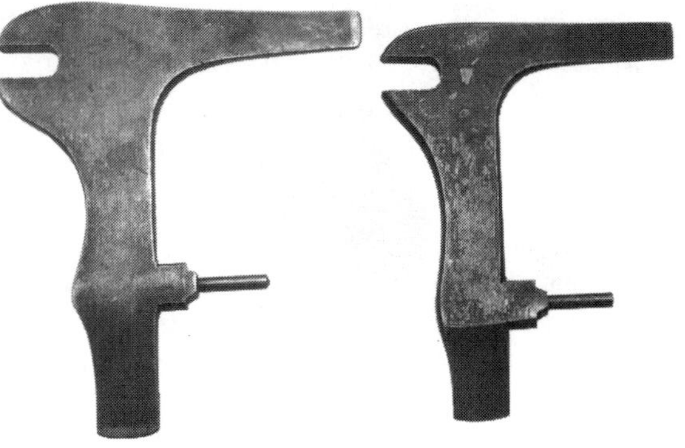

Fig. 19-25. The larger-size Combination Tool, Experimental, 1870 (left), was issued with the rifles and carbines used in the 1871–1872 Field Trials. The smaller version (right) was issued in 1874 and later for use with the U.S. Model 1873 .45-70 Springfield arms.

In fact, the smaller variation probably was not issued at all during the 1871–72 trial period, but sometime after 1874. Craig Riesch has pointed out that the confusion may have arisen when collectors improperly compared the Combination Screw Driver, Experimental, 1870 pictured in the *Description and Rules for the Management of the Springfield, Remington, and Sharp's Rifle Muskets, Experimental Models, 1870*, published in 1871 and *Description and Rules for the Management of the Ward-Burton Rifle Musket, Model 1871*, published a year later in 1872. The Combination Screw Driver pictured in the former is clearly a roughly drawn sketch while that in the Ward-Burton manual is a precision artist's engraving cut produced specifically for printing and is drawn to scale. The text in the Ward-Burton manual clearly states that the:

> "Combination Screw Driver,—shown in the cut below, is used in common by all the trial arms."

That alone should tell us the Ward-Burton was issued with the same screwdriver as the other trial arms, which was already being produced.

Additionally, Mr. Riesch points out that the Combination Screw Driver shown in the Ward-Burton manual has the "spring vise" opening for a "V"-shaped mainspring and a tumbler punch. The Ward-Burton does not have either a "V" mainspring or a hammer tumbler. If the Springfield Armory intended to manufacture a special tool for the 1,013 Ward-Burton rifles and 317 Ward-Burton carbines, why go to the additional cost of machining unneeded features?

Finally, the Combination Screw Driver pictured in the *Description and Rules for the Management of the Springfield Rifle, Carbine, and Army Revolvers, Calibre .45* published two years later in 1874 is the same cut as used in the Ward-Burton manual, i.e., exact same shape and size, see Figure 19-26. The 1874 manual states:

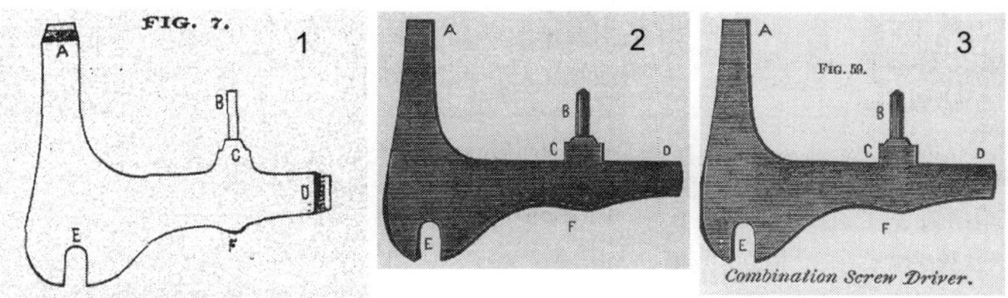

Fig. 19-26. These illustrations were scanned 1:1 directly from (1) *Description and Rules for the Management of the Springfield, Remington, and Sharp's Rifle Muskets* . . . 1871; (2) *Description and Rules for the Management of the Ward-Burton Rifle Musket, Model 1871* . . . 1872; and (3) *Description and Rules for the Management of the Springfield Rifle* . . . 1874. (1) is a drawing and (2) and (3) are clearly the same printers' cut and are exactly the same size in both manuals.

"Combination Screw Driver,—[shown] two thirds size—."

When the differences in the sizes of the respective cuts are measured and the ratios calculated, they are in fact the same size as the larger Combination Screw Driver which is 2.95 inches high by 2.43 inches wide. *All* of the trials rifles and carbines used the larger Combination Screw Driver.

The larger tool is the same shape in all three manuals; it is the one issued for all arms during the period of the Field Trials and proposed for the Model 1873 rifle and carbine. The smaller Combination Screw Driver, as seen in Figure 19-25, has a different shape and is smaller in size at 2.67 inches high by 1.95 inches wide. This tool was apparently not described in any Ordnance Department manual, or if so, its description has not yet been located. The measurements of this smaller screwdriver are close to the measurements of the Model 1879 Combination Tool, and it is believed to be a transitional tool dating from circa 1874–75 to the final adoption of the Model 1879 Combination Tool.

Table 19-2 presents the dimensions of the two experimental combination tools.

Table 19-2
Experimental Combination Tool*

Date of Issue	Overall Height	Overall Width
1871	2.95 inches	2.43 inches
1874-1877**	2.67 inches	1.95 inches

* No Ordnance Department official designation with a year of issue appears to have been given to these tools.
** No record of this tool's existence or period of issue has yet been discovered, but examples are known to exist including one in the C. Riesch collection.

Model 1871 Screwdriver for the U.S. Model 1871 (Springfield-Spencer) Infantry Rifle

Because a separate manual was not written for the U.S. Model 1871 "Infantry Rifle" (as altered from the Model 1865 Spencer Carbine), we have to guess at which tool, if any, was packed in the rifle shipping crates. The Spencer carbines altered to .50 caliber after the Civil War were issued the combination tool that had been used during the Civil War, a two-bladed screwdriver with the blades connected by a rivet, see Figure 19-27. The long blade had a small hole punched into its top for a thong. Some authorities suggest that these screwdrivers were shortened for post–Civil War issue, others that new screwdrivers were manufactured with a shorter main leg. Still others suggest that two sizes had been issued during the war, one long for the rifle and the other short for the carbine. And a fourth faction suggests that both sizes were manufactured, probably by different subcontractors.

Fig. 19-27. It is likely that this post–Civil War variation of the Spencer screwdriver was packed in the U.S. Model 1871 Springfield-Spencer shipping crates.

Whichever theory is correct, we do suspect that only the short screwdriver, if any, was issued for the Spencer in the postwar period. The main blade was 0.5 inch wide and 2.9 inches long; the short blade was 0.55 inch wide and 2.25 inches long.

Chapter 20: The U.S. Bayonet, 1865–1872

U.S. Model 1855 Bayonet and Scabbard

The standard musket bayonet in use at the beginning of our time period was adopted in 1855. It had an 18-inch blade and was triangular in cross section. The blade was nominally 0.77 inch wide at the widest point. The socket was 3 inches long and its internal diameter was nominally 0.820 inch at the rear, tapering to 0.782 inch in diameter at the front. The entire bayonet was finished National Armory Bright, see Figure 20-1.

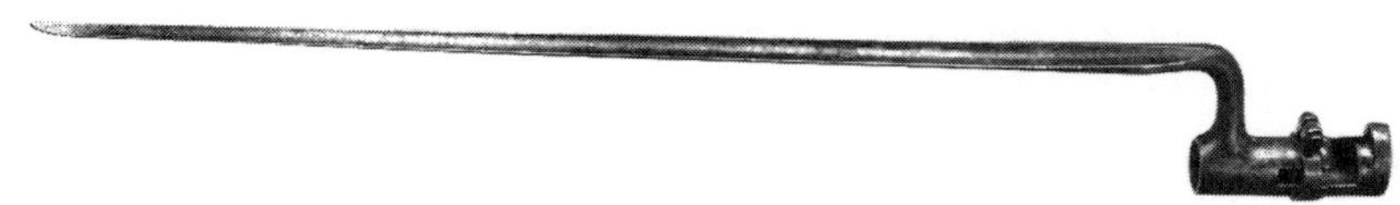

Fig. 20-1. Model 1855 Bayonet, as used on all the U.S. Model rifles and trial rifles from 1865 through 1870. Thomas Fleming collection.

Scabbards issued to the troops for the Model 1855 Bayonet were varied. The earliest type associated with the rifles under study were simply Civil War surplus, having leather bodies and a wide, curved belt loop, see Figure 20-2. An intermediate style retained the leather body but had a swiveling frog, held by a cast brass escutcheon approximately 1.08 inches in diameter, see Figure 20-3. The frogs appear

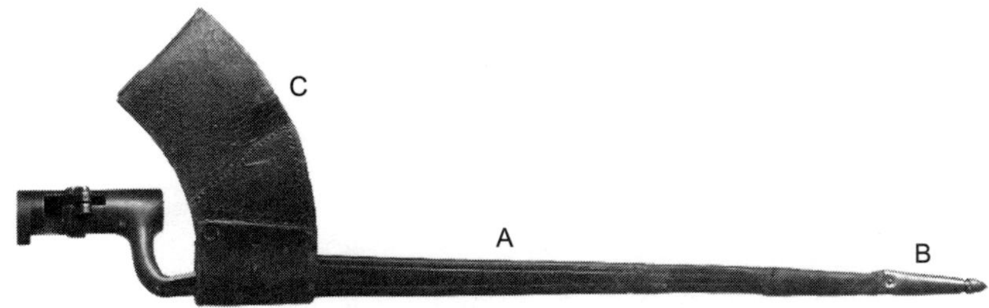

Fig. 20-2. Early Model 1855 Civil War–style leather bayonet scabbard (A) with brass chape (B) and a large curved frog (C) for attachment to waist belt.

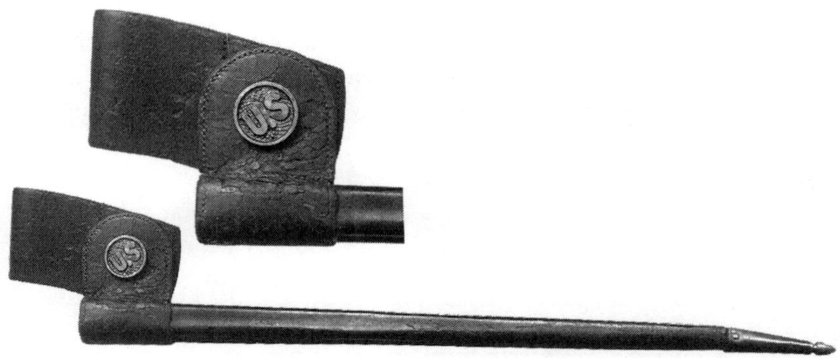

Fig. 20-3. Later, intermediate-style scabbard (scarce) with Hoffman swivel frog. This model retains the leather body and brass chape of the earlier Model 1855 scabbard. Inset shows frog in greater detail. Notice the "S" lacks a period. Thomas Fleming collection.

in several styles and vary in stitching, riveting, and cut. Escutcheons on the scabbards used during this time have a single period between the "U" and "S" but no period after the second letter. The brass escutcheons without periods after either letter appeared after 1874. In 1868, the scabbard body was changed from leather to sheet iron and continued in use without change with the U.S. Model 1873 and 1884 .45-70 Springfield rifles.

All Model 1855 bayonet blades will fit all (full-length) scabbards. This is not surprising as all Model 1855 bayonet blades are identical. In fact, after 1877, Model 1873 .45-70 bayonets are really Model 1855 bayonets which were subjected to a cold-pressing process whereby the sockets were squeezed down to fit the smaller barrels. The "fins" of extruded metal were ground off and the bayonet polished and browned, see Figure 20-4.

By 1890, the Springfield Armory had ex-

Fig. 20-4. Model 1855 (top) and Model 1873 bayonet (bottom) socket diameters compared.

hausted its supply of serviceable Model 1855 bayonets. Rather than tool up to produce entirely new Model 1873 bayonets which were seen as becoming obsolete anyway, it was decided to adopt the combination ramrod and bayonet which was used to the end of Model 1888 rifle production.

Table 20-1
Bayonets in Use, 1865–1872, by Arm

Rifle	Bayonet
U.S. Model 1865 .58 Caliber Springfield Rifle	Model 1855 (unmodified)
U.S. Model 1865 .58 Caliber Springfield Cadet Rifle	Model 1855 (may have been shortened and the blade narrowed but no Model 1865 Cadet bayonets have yet been positively identified)
U.S. Model 1866 .50-70 Caliber Springfield Rifle	Model 1855 (unmodified)
U.S. Model 1866 .50-55 Caliber Springfield Cadet Rifle	Model 1855-type, 2-15/16-inch-long socket with 0.745-inch ID. Blade 11/16 inch (0.69 inch) wide, 16-1/4 inches long on most specimens (similar bayonet: 14-1/2-inch blade, 3-inch socket). Specimens encountered which appear newly made. Examples bushed to these dimensions are also known.
U.S. Model 1868 .50-70 Caliber Springfield Rifle	Model 1855 (unmodified)
	An estimated 704 Model 1868/69 Trowel (or entrenching) Bayonets were manufactured in two batches of 200 and 504. Made by welding a blade 3-7/8 inches wide by 8-15/16 inches long to a Model 1855 socket

U.S. Model 1869 .50-55 Springfield Cadet Rifle	Model 1855-type, 2-15/16-inch-long socket with 0.745-inch ID. Blade 11/16 inch (0.69 inch) wide, 16-1/4 inches long on most specimens (similar bayonet: 14-1/2-inch blade, 3-inch socket). Specimens encountered which appear newly made. Examples bushed to these dimensions are also known.
U.S. Model 1868 Springfield-Remington (.50-70) Transformed Rifle	Model 1855 (unmodified)
U.S. Model 1870 .50-70 Caliber Springfield Rifle	Model 1855 (unmodified)
U.S. Model 1867.50-45 Navy Cadet Rifle (Remington Patents)	Model 1855-type, 2-15/16-inch-long socket with 0.745-inch ID. Blade 11/16 inch (0.69 inch) wide, 16-1/4 inches long on most specimens (similar bayonet: 14-1/2-inch blade, 3-inch socket). Specimens encountered which appear newly made. Examples bushed to these dimensions are also known.
U.S. Model 1870 Navy Rifles, Type I and Type II	Sword Bayonet: cast brass grip 4.25 inches long; blade 20 inches long, one stopped fuller 14-5/8 inches long, spear point.
U.S. Model 1870 Trials Rifle (Remington Patents)	Model 1855 (unmodified)
U.S. Model 1871 Army Rifle (Remington Patents)	Model 1855 (unmodified)
U.S. Model 1871 Infantry Rifle as altered from the Model 1865 Spencer Carbine	Model 1855 (unmodified)
U.S. Model Sharps Rifle Musket, Experimental Model, 1870, both types.	Model 1855 (unmodified)
U.S. Model Ward-Burton Rifle Musket, Experimental Model, 1871	Model 1855 (unmodified)

The .58- and .50-Caliber Rifles

U.S. Model 1855 Bayonet Issues

U.S. Model 1865 .58 Caliber Springfield Rifle Bayonet

The Model 1865 .58 Caliber Rifle used the standard Model 1855 bayonet for which it was equipped in its original configuration as a percussion rifle musket. In fact, this same bayonet was used, without alteration, on the entire production of Springfield arms, standard and experimental/trial, except for the various "Cadet" models, made prior to the .45-70 U.S. Model 1873 Springfield.

U.S. Model 1866 .50-70 Caliber Springfield Rifle Bayonet

The Model 1855 bayonet in its metal-tipped leather scabbard was issued for this rifle without modification.

U.S. Model 1866 (1867) .50-55 Caliber Springfield Cadet Rifle Bayonet

The bayonet provided with the Model 1866 Cadet Rifle was similar to the Model 1855 bayonet, but was slightly smaller, and was apparently newly manufactured as opposed to an alteration of existing Model 1855 bayonets, though some examples of the latter are known with bushed sockets.

The socket was bored to approximately 0.745 inch to fit the smaller muzzle diameter. The bayonet was polished bright and the face flute was stamped "U.S." The blade was narrower, at 11/16 inch (0.69 inch), and shorter, at 16-1/4 inches. The scabbard length was proportional to the bayonet.

U.S. Model 1868 .50-70 Springfield Rifle Bayonet

The U.S. Model 1868 Rifle was issued with a surplus M1855 socket bayonet, finished bright, from arsenal stockpiles.

U.S. Model 1869 .50-55 Springfield Cadet Rifle Bayonet

The bayonet provided with the Model 1869 Cadet Rifle was similar to the Model 1855 rifle musket bayonet, but was slightly smaller, and was apparently newly manufactured as opposed to an alteration of existing Model 1855 bayonets, though some examples of the latter are known with bushed sockets. See Figure 20-5. Also see comment for Model 1866 (1867) as above.

Fig. 20-5. The bayonet for the U.S. Model 1869 .50-55 Springfield Cadet Rifle was a newly manufactured bayonet although some modified versions of the Model 1855 do exist.

The socket was bored to approximately 0.745 inch to fit the smaller muzzle diameter. The bayonet was polished bright and the face of the flute was stamped "U.S." The blade was narrower, at 11/16 inch (0.69 inch), and shorter, at 16-1/4 inches. The scabbard length was proportional to the blade.

U.S. Model 1868 Springfield-Remington (.50-70) Transformed Rifle Bayonet

The Model 1855 socket bayonet and scabbard were issued with this converted rifle. The bayonets were finished bright.

U.S. Model 1867 .50-45 Navy Cadet Rifle (Remington Patents) Bayonet

The Model 1866 Cadet Bayonet was used for this model. It was apparently a newly manufactured bayonet although it appears that some existing Model 1855 bayonets were modified. Also, see the description for the Model 1866 (1867) Cadet, above.

The socket was bored to approximately 0.745 inch to fit the smaller muzzle diameter. The bayonet was polished bright and the face flute was stamped "U.S." The blade was narrower, at 11/16 inch (0.69

inch), and shorter, at 16-1/4 inches. The scabbard length was proportional to the bayonet.

U.S. Model 1871 Army Rifle (Remington Patents)

U.S. Model 1855 bayonet and scabbard, unmodified, were issued with this rifle.

U.S. Model 1871 Infantry Rifle (Spencer Alteration)

U.S. Model 1855 bayonet and scabbard, unmodified, were issued with this rifle.

Bayonets for the Experimental Rifles Issued for the 1871–1872 Trials

The four rifles (Springfield Model 1870, Sharps Model 1870, Remington Model 1870, and the Ward-Burton Model 1871) were all equipped with the standard U.S. Model 1855 bayonet and scabbard, unmodified.

Other Bayonet Models Issued, 1865–1872

U.S. Model 1868 Trowel Bayonet

During the period when the U.S. Model 1868 and 1870 .50-70 rifles were in use, field tests were conducted with two very similar models of trowel—or entrenching—bayonets, following the design of Edmund Rice. The intent was to provide the soldier with a tool which not only could be used to create a sheltered firing position, but which could also serve as a weapon.

Approximately 704 (possibly less, depending on how the records are interpreted) trowel bayonets were produced, from 1869 to 1871 in two models, that of 1868 and 1869. Both were finished bright, and furnished with leather and brass scabbards with a belt loop in the form of a sweeping curve. Both types of trowel bayonet were fabricated by welding a wide, short, pointed "digging" blade (approximately 3-7/8 inches wide by 8-15/16 inches long) to a Model 1855 socket bayonet. A third variation was the Model 1873, a much-improved version by

FelixChillingworth, in which a new blade was welded to the ground-down blade of a Model 1855 bayonet. The ground blade of the Model 1855 provided a stiffening "spine" for the new bayonet.

Fig. 20-6. The U.S. Model 1868 Trowel Bayonet was modified from the U.S. Model 1855 Bayonet. It was finished bright and held in a leather and brass scabbard. Note the finger loop (arrow).

Only the Model 1868 Trowel or Entrenching Bayonet is of concern in the period 1868–1872. It had a simple looped finger guard (see Figure 20-6), while the Model 1869 Entrenching Bayonet incorporated, as part of the loop, a pivoting felt-faced tompion to close off and protect the muzzle. The Trowel Bayonet was to be used to dig entrenchments which at the time were little more than shallow depressions with the earth heaped up in front a foot or so high to protect the soldier while firing from the prone position. The troops were strongly admonished that the bayonet was absolutely not to be used for digging while attached to the rifle! Even so, this remained a major concern of the Ordnance Department.

The exact production breakdown, by type, is unknown since extant records are confusing, even conflicting, but both models are considered very scarce and desirable items today; the Model 1869 with the attached tompion, especially so. Field trials produced mixed reviews. It was found to be very effective in loose dirt, but the handle design was quite uncomfortable to use when digging in hard ground. Fears persisted that damage to the rifle barrel would occur if the soldier elected to use his rifle as a shovel handle. Neither pattern was adopted for general issue.

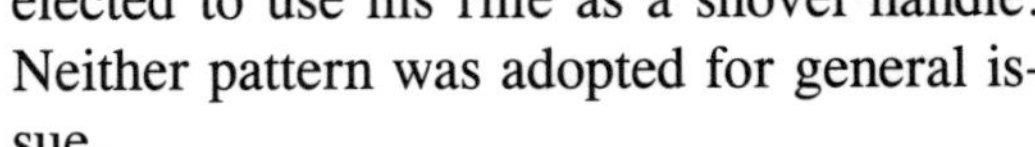

A later 1873 version was designed by Felix Chillingworth, and 10,000 were produced during 1876 for use with the Model 1873 Rifle. The Model 1873 Trowel Bayonet remained in service until 1890 when it was withdrawn due to the adoption of the U.S. Model 1888 .45-70 Rod Bayonet Rifle.

U.S. Model 1870 Navy Rifle, Type I and Type II, Bayonet

Unlike any other Springfield arm of this period, but in keeping with previous Navy tradition, the U.S. Navy Rifles, both Type I and Type II, mounted sword bayonets, rather than the angular Model 1855 bayonet, see Figure 20-7. Most such bayonets were mounted on the right side of the barrel but in the Model 1870 Navy rifles, they were mounted beneath the barrel. This required that the cast brass hilt be bored out, to slip over the ramrod.

The bayonets were manufactured by the Ames Manufacturing Company of Chicopee, Mass. Two variations are known and distinguished by blade shape: the Type I had a shallow "S" or "yataghan" profile. The Type II, which was produced in far greater quantity, had a straight blade. Both hilts had an awkward and uncomfortable

Fig. 20-7. U.S. Model 1870 Navy Rifle, Type II Sword Bayonet and scabbard. Roy Marcot collection.

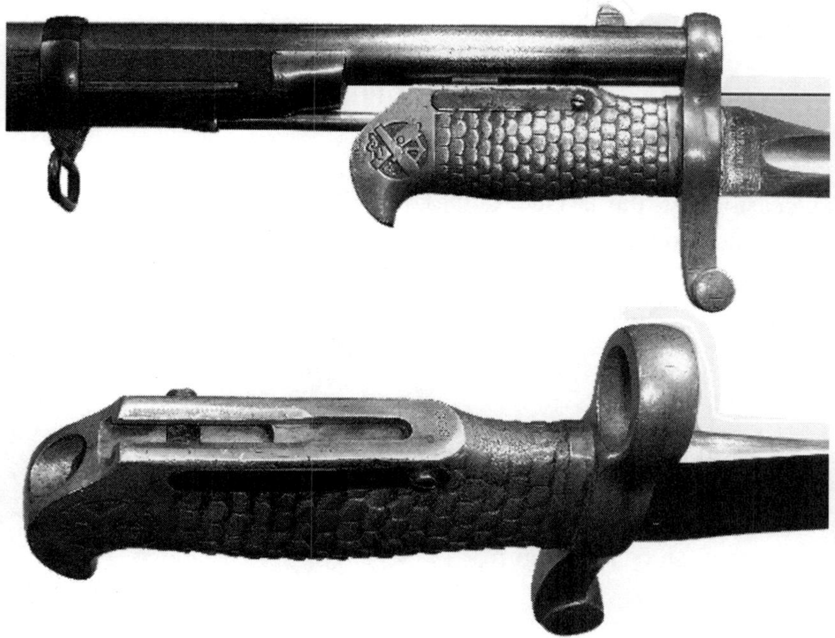

Fig. 20-8. These views show details of the cast brass grip and spring catch mechanism. The bayonet slides over the ramrod. All other U.S. sword bayonets mount on the right side of the barrel.

grip, because the catch mechanism was forced to protrude above the top surface to allow room for the ramrod, see Figure 20-8.

The cast brass grip was 4.25 inches long, with a pronounced "beak" pommel. There were seventeen rows of scales on either side of the grip. The Navy Bureau of Ordnance escutcheon—crossed cannons with fouled anchor—was molded into the pommel at the rear, just above the beak, on both sides. The cross guard was 0.50 inch thick by 3.82 inches wide. The top arm had a loop with a 0.770-inch bore, while the lower tang ended in a flat ball finial. The blade was 20 inches long, with a 14-5/8-inch-long single, stopped, fuller centered on each side. It tapered symmetrically from 0.32 inch by 1.145 inches at the guard, to a spear point. Both edges were sharpened ("V" ground) for approximately the first 15 inches from the tip.

Following established convention, markings are located on the obverse ricasso:

AMES MFG CO.
CHICOPEE
MASS.

Occasionally, the last word, "MASS.," will not be visible. The reverse ricasso was marked:

U.S.N.
G.G.S.
1870

Fig. 20-9. The Navy inspector's initials are stamped on the reverse of the ricasso on the U.S. Model 1870 Navy bayonet.

The inspector's initials, "G.G.S.," are repeated on the brass hilt, before the mounting slot, see Figure 20-9.

NOTE: A few years ago, Model 1870 Remington Navy "Sword Bayonets" were offered for sale by a surplus arms dealer. They were purported to have an "original" blade with a cast reproduction of the "Navy-style" hilt. The blade does *appear* to be original but devoid of any marking. Examination of three samples suggests that any original markings were ground off a long time ago.

The minimal workmanship on the Bureau of Ordnance Escutcheon on the pommel also readily identifies it as a reproduction. When tried on an original U.S. Model 1870, Type II Navy rifle, the bayonet did fit almost as well as an original. A reproduction scabbard was also available about the same time from another source. Again, the workmanship appeared quite good until compared to an original scabbard. See Figure 20-10 for a photograph of the bayonet and scabbard and Figure 20-11 for detail of the Navy Bureau of Ordnance Escutcheon.

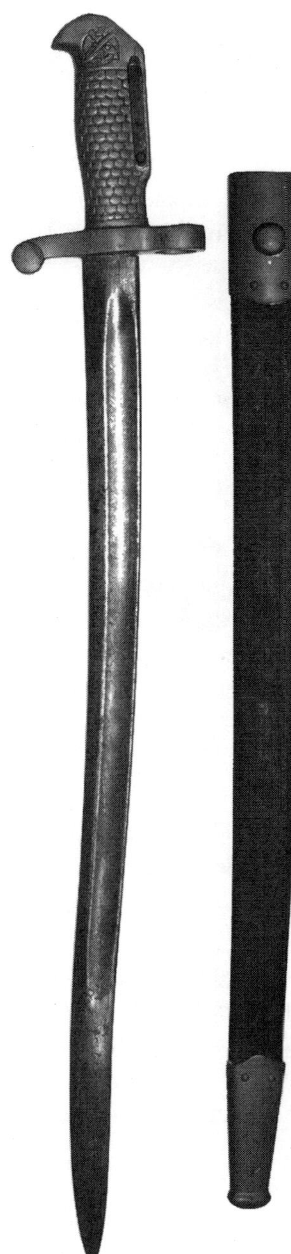

Fig. 20-10. This example of a Model 1870 Navy Type I bayonet is a reproduction which used a possibly original blade with a newly cast brass hilt. The scabbard is also a reproduction. They were produced in the early 1990s.

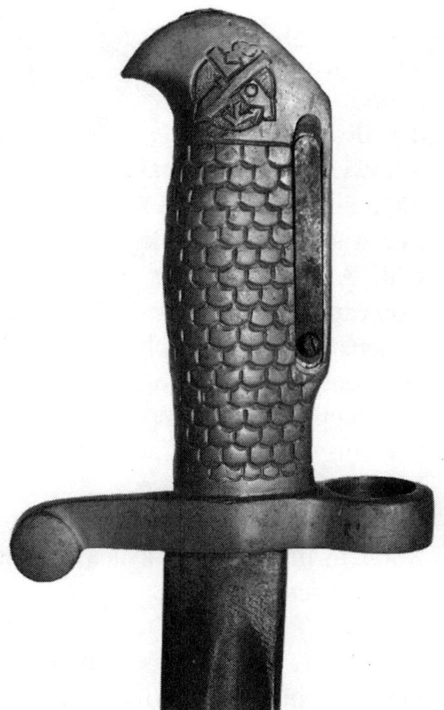

Fig. 20-11. This close-up of the hilt shows the crude rendition of the Navy's Bureau of Ordnance escutcheon on the pommel. Compare it to Figure 20-8. Note also the absence of the Ames markings on the ricasso.

Chapter 21: Ammunition for the .58- and .50-Caliber Rifles and Carbines of Springfield Armory, 1865–1872

Introduction

The Civil War, at least from the infantry standpoint, had been fought mainly with muzzleloading rifled muskets, mostly of .58 caliber, using preassembled paper cartridges, see Figure 21-1. The soldier drew the cartridge from his ammunition box, bit or tore open the end and poured the loose powder down the muzzle, followed (sometimes) by the cartridge paper as a wad, and the Minié ball. A copper percussion cap was taken from a separate cap pouch and pushed down on the nipple. This was a clumsy procedure at best and was far more difficult while in motion, or under fire. Tactics still dictated massed troop formations and a heavy reliance on volley fire by ranks, all under strict command.

Fig. 21-1. The .58-caliber paper cartridge of the rifle-musket era was quickly superseded in U.S. service by the metallic self-contained cartridge following the end of the Civil War.

There were other perils involved with muzzleloading. A lost ramrod meant a rifle musket out of action. Under severe stress of combat, men could (and did) load one round on top of another without actually firing. If and when they did fire, the result was at best a blown rifle barrel, or at worst a needless casualty. The breechloader firing metallic cartridges eliminated both problems.

Union cavalry units were armed with a wide variety of breechloading carbines using an equally wide assortment of patented self-contained or partially self-contained cartridges. Some varieties such as the Spencer and Henry rimfire cartridges were very practical, while others like the paper or linen cartridges used in the Sharps, or those with rubber cases for the Smith, were too delicate for field use and had high rates of wastage.

The primitive metal cartridges for the Burnside, Gallagher, Maynard, etc., were not much better than paper, as they still required the use of a separate percussion cap, and the arms lacked fully functional extraction/ejection mechanisms.

The following paragraphs discuss those metallic, self-contained cartridges used in the Springfield breechloading arms that were described in preceding chapters.

NOTE: An interesting and increasingly valuable group of artifacts are the original packets of ammunition. The survival rate of these low-quality pasteboard boxes, containing up to twenty heavy, poorly packed cartridges with only a thin paper overwrap, is very low. The .50-caliber boxes are much scarcer than their .45-caliber counterparts, both because they are older and also because fewer were originally made.

.50-60 Joslyn Rimfire Cartridge

M1865 Springfield-Joslyn Rifle, Type I (Chapter 10)

This cartridge was originally developed for the Model 1865 Springfield-Joslyn rifles described in Chapter 10, see Figure 21-2. Work on the new cartridge began in March 1865 to develop a .50-caliber rimfire cartridge with a 60-grain charge and a 400-grain bullet. The overall length of the car-

Fig. 21-2. The .50-60 Joslyn cartridge was developed for the Model 1865 Springfield-Joslyn. Left, cartridge manufactured at the Frankford Arsenal; right, cartridge manufactured in France. Lou Behling collection.

tridge was 1.93 inches and the copper cartridge case was 1.47 inches long. The cartridge case was not headstamped.

On March 24, 1865, General A.B. Dyer sent a telegram to Silas Crispin, at Frankford Arsenal, stating: "Stop the manufacture of the metallic primed cartridges for the .50 caliber musket, samples of which were furnished by Maj. Laidley. The new model carbine cartridge will take its place." The "new model carbine cartridge" was, of course, the .56-.50 Spencer cartridge. Today, the original cartridge is better known as the .50-60 Peabody.

Supplies of this cartridge were small, and few have survived. The .56-.50 Spencer cartridge was issued as a substitute. Since the Spencer cartridge contained only 45 grains of black powder, some have considered it analogous to firing .22 Shorts in a .22 Long Rifle chamber. The author is indebted to Lou Behling for this information.

.56-.50 Spencer Rimfire Cartridge

M1865 Springfield-Joslyn Rifle, Type I (Chapter 10) (alternate)
M1871 Springfield-Spencer Conversion (Chapter 16)

The .56-.50 Spencer was developed during 1863–1864 as an improvement over the original .56-.56 rimfire cartridge. The Spencer cartridge was somewhat unique in its nomenclature, see Figure 21-3. The first number was the diameter of the case just ahead of the rim, and the second, the diameter of the case mouth. The .56-.50 had a slightly tapered case while the original cartridge had a straight case. The slight taper made it easier to extract the fired case from the breech. It was only 1.63 inches long overall and its 1.17-inch-long copper case was filled with a relatively small charge of 45 grains of black powder. The lead bullet was conical and weighed 350 grains.

Fig. 21-3. Spencer rimfire cartridges. Left to right: .56-.56, .56-.52, and .56-.50.

This cartridge did not see Civil War service but was introduced starting in mid- to late 1866. Returns from the Second Quarter of 1867 show that it had already been issued to units of the 2nd, 3rd, 4th, 6th, 7th, 8th, 9th, and 10th Cavalry Regiments. It was used most effectively at the Battle of Beecher's Island, Nebraska Territory, in 1867.

A second Spencer cartridge having a head diameter of .56 inch, a case mouth of 0.52 inch (.56-.52) and a slightly shorter case was, for all intents and purposes, interchangeable with the .56-.50 cartridge.

The Spencer cartridge was manufactured by a number of different ammunition companies and so is found in cardboard boxes of varying designs and capacities, from ten to forty-two rounds per box. The latter number of cartridges exactly filled the six-tube Blakeslee Patent Spencer carbine box, described in Chapter 19.

The .56-.56 Spencer cartridge had performed well for the Union in the short-range skirmishes that marked cavalry tactics during the Civil War. And from 1866 through 1870, Spencer carbines firing the .56-.50 cartridge had served the mounted troops, but that cartridge, in either the postwar .56-.52 or .56-.50 chamberings, lacked the range required on the plains and deserts of the western territories,

The U.S. Model 1871 "Infantry Rifles" (as altered from the Model 1865 Spencer Carbine) and described in Chapter 16, were never issued. If they had been, they would have used this ammunition.

By the early 1870s, the Spencer carbine had been withdrawn from service with the regular army. But it continued to be issued for several years to civilian contractors such as teamsters and Indian scouts.

U.S. Model .58-60 Cartridge

M1865 Springfield-Allin Rifles (Chapters 1 and 2)

The U.S. Model .58-60 cartridge was used in the U.S. Model 1865 Springfield Rifle (First Allin Alteration), both the service and cadet rifles, see Figure 21-4. A rimfire cartridge, it was of .58 caliber, charged with 60 grains of black powder, and fired a 500-grain lead bullet. The bullet diameter was, of course, selected so that existing rifle musket barrels could be used without modification. The 60-grain powder charge was the same as had been used in the U.S. Model 1855 through 1864 rifle muskets. The ballistics of this cartridge were rather limited. The muzzle velocity has been calculated at about 1,040 feet per second (fps) and the muzzle energy at 1,200 foot-pounds (ft-lb) (Butler, *United States Firearms, The First Century, 1776-1875*).

The Ordnance Department was not satisfied at all with the use of the rimfire system for a service cartridge as this paragraph from *Ordnance Memoranda No. 14* makes clear:

> "In [the rimfire cartridge case] the fulminate composition was placed in the folded head of the case. This mode of priming requires a large charge of the priming composition, which being thrown into the fold by swiveling , the entire circumference of the head was not always primed thoroughly, and as the cartridge is exploded by striking the rim at a part of the head under the hammer, it not infrequently . . . failed. The large charge required (about 5 grains against 1/2 grain for the centerfire), was a further objection . . . the exploding of so large a quantity

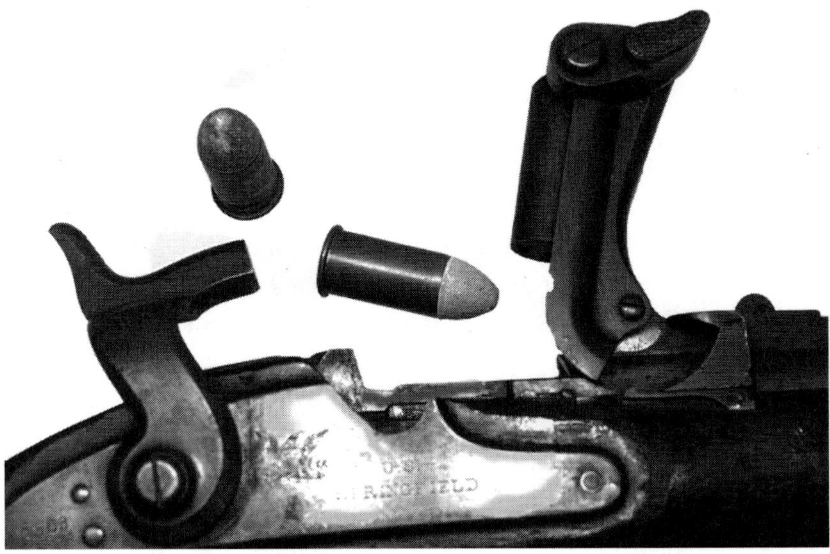

Fig. 21-4. U.S. Model 1865 .58-60 cartridge.

> of . . . (priming) . . . in the folded-head, the weak part of the cartridge, tended to strain and open the fold to bursting . . . another objection . . . is that they are more liable to accident in handling, and in shocks of transportation, and those incident to service."

Great strides in metallurgy, and especially in the drawing of cartridge cases, were occurring at the time and it quickly became possible to not only reduce the standard caliber to 0.50 inch but to manufacture a cartridge case long enough to hold a 70-grain charge, large enough to provide a muzzle velocity of 1,240 fps with a muzzle energy of 1,400 foot-pounds.

U.S. Model .50-70-450 Cartridge

Model 1866, Model 1868, Model 1870 Springfield Rifles (Chapters 3, 6, 8)
Model 1865 Springfield-Joslyn Rifle, Type II (converted from rimfire) (Chapter 10)
Model 1868 Springfield-Remington Transformed Rifle (Chapter 11)
Model 1870 Remington Navy Rifle, Types 1 and II (Chapter 13)
Model 1870 Remington Rifle Musket, Experimental (Chapter 14)
Model 1871 Army Rifle (Remington Patents) (Chapter 15)
Model 1870 Sharps Rifle Musket, Experimental Model, 1870, Type I and II; Carbine, Type 1 (Chapter 17)
Model 1871 Ward-Burton Rifle Musket, Experimental Model Chapter 18)

An entire book could be devoted to the .50-70 cartridge and its shorter cousins. There were countless major and minor variations in primer type and construction, case material including the strange (and rare) coiled brass version, using the Rodman-Crispin patent, various bullet shapes, etc.

The Frankford Arsenal began development of a centerfire cartridge as early as 1864. Using rimfire blanks, they attached an anvil which was in the shape of a flat metal plate. The plate was indented into a cup with a center hole to hold a percussion cap with the open end pointing toward the mouth of the case. The plate was slightly smaller in diameter than the cartridge case. When the case was charged and the bullet seated, the plate was held in place against the thin copper head. When the hammer struck the center of the case head, it crushed it against the anvil to ignite the percussion cap; the hot gases blew through the hole in the anvil and ignited the powder charge.

The concept worked and improvements followed quickly. The cartridge case was crimped to hold the anvil securely in place; then the anvil was narrowed to a "bridge" of metal that held the priming compound in a specially made cup that proved more efficient than a standard percussion cap. Still other cartridges were developed in which the case head was indented to form a pocket which held the priming compound so that it poured into the case and was secured with a perforated metal plate which served as the anvil.

The new .50-inch-diameter cartridge was manufactured in several lengths. The blank was normally 1.522 inches long. The cases were formed from disks of copper which were punched, or deformed in the center in a series of dies to form a tube with one open end. The bottom of the case was further deformed, or folded, to form the rimmed head.

The first successful centerfire design adopted for U.S. military service was the Frankford Arsenal "Bar Anvil" cartridge developed by E.H. Martin in 1866 at the Springfield National Armory, see Figure 21-5. A rigid tin anvil cup in the form of a three-sided hollow bar was formed which fit inside the copper cartridge case. The priming compound was contained in a recess in the center of the cup. The sides of the cartridge case just below the head and outside the end of the bar were crimped in a circular manner to hold the cup in place. Major T.J. Treadwell reported in his treatise, *Metallic Cartridges (Regulation and Experimental) as Manufactured and Tested at the Frankford Arsenal, Philadelphia, PA., 1873*, that several million cartridges were manufactured and issued between October 1866 and March 1868.

The bar anvil was superseded in 1868 by an improved design developed by Colonel Stephen Benét which used a thicker-walled, round copper cup which was held in place by two crimps above the cup, see Figure 21-6. Two holes were punched through the cup for the flame from the priming compound. This virtually eliminated the problem of the bar anvil blowing loose into the barrel, or of being inserted upside down. Occasional misfires did occur, however, because of the thicker case head. This design, with a 450-grain lead bullet, became the standard service cartridge, and remained in use well beyond 1872; it was also adopted for initial production of the U.S. Model 1873 .45-70 service round, lasting until the adoption of the Boxer-primed cartridge case in 1882.

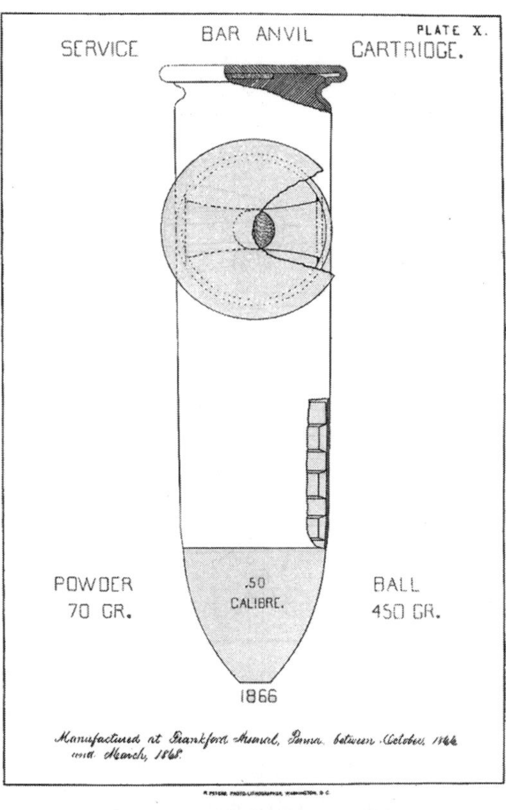

Fig. 21-5. Drawing of the Frankford Arsenal .50-70 Bar Anvil cartridge in use between October 1866 to March 1868. Major T. J. Treadwell, Ord. Dept.

The .50-70 cartridges intended for U.S. military service were manufactured at the Frankford Arsenal (see Figure 21-7) or by subcontractors such as the U.S. Cartridge Company of Lowell, Massachusetts. They were packed in pasteboard boxes of twenty cartridges. The boxes were imprinted with the name of the Arsenal, the caliber, load and bullet weight of the cartridge, and the date of manufacture. Lot numbers were not used during this period. Figure 21-8 shows a complete package. Note the dates stamped on the box.

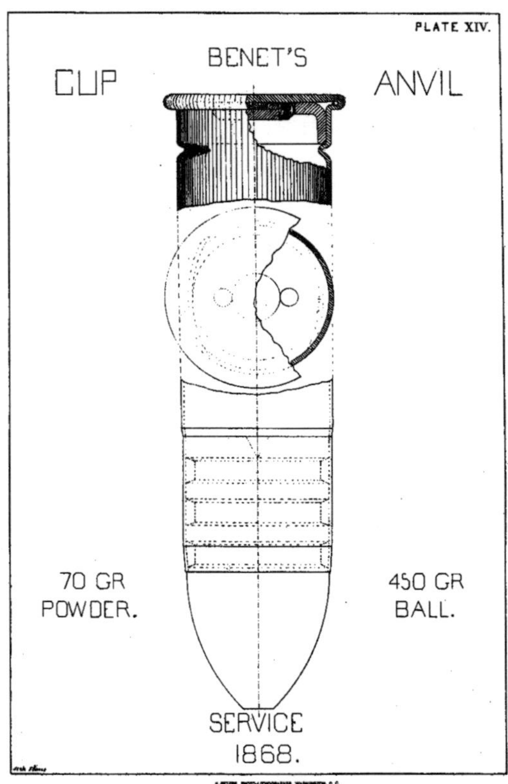

Fig. 21-6. Drawing of the Frankford Arsenal .50-70 Benet Cup Anvil cartridge in use from 1868 to 1882. Major T. J. Treadwell, Ord. Dept.

With a single exception of one lot of .50-70 Benet-primed cartridges in September of 1882 (R F 9 82) none of the early inside-primed ammunition was headstamped. There were a few exceptions, but full-scale headstamping of cartridges did not begin at the Frankford Arsenal until 1877.

After Frankford Arsenal switched its primary production effort to the .45-70 round in 1874, the government purchased many .50-70 cartridges under contract; one large supplier was the United States Cartridge Company of Lowell, Massachusetts, see Figure 21-9. These later versions of the .50-70 used brass cases, external primers (probably Farrington-primed which was the then-current standard at the company) and were therefore reloadable. Much of the later contract ammunition, especially in the .45-70 period, bore the manufacturers' normal commercial headstamp.

This style of box is pictured in the Pitman Notes, Volume 5. It is constructed in the same manner as the military boxes made at Frankford Arsenal, from pasteboard sections assembled into a box which formed a parallelogram in cross-section with notched interlocking spaces, nine short and one long, the whole wrapped with paper, having an embedded tear string. The dimensions are slightly smaller at 6-3/16

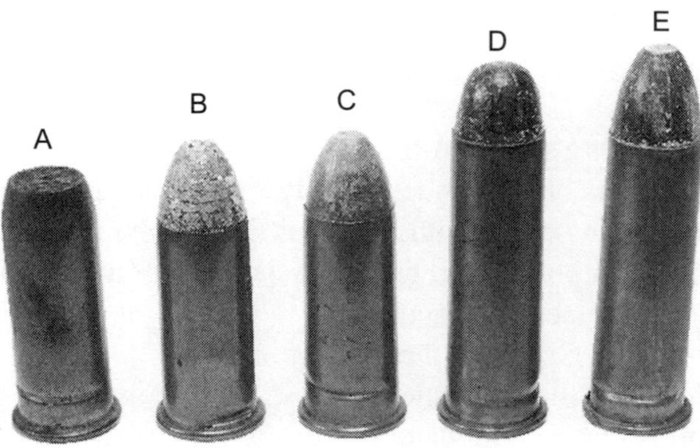

Fig. 21-7. These .50-70, .50-55, and .50-45 cartridges were manufactured at the Frankford Arsenal. Left to right: (A) Blank; (B) Martin or Bar Anvil-primed (Navy) Cadet; (C) Benet-primed Cadet; (D) Benet-primed Carbine; and (E) Benet-primed Service Rifle.

Fig. 21-8. An example of a box of .50-70 Frankford Arsenal–manufactured cartridges packaged in 1871. This label style was used through 1873.

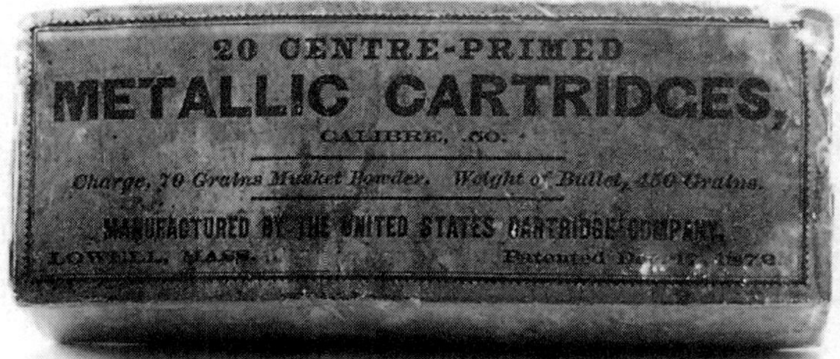

Fig. 21-9. These .50-70 cartridges were manufactured under contract to the Ordnance Department by the U.S. Cartridge Company of Lowell, Mass., in the early 1880s.

inches x 2-5/16 inches x 1-3/8 inches thick. The label is printed in black ink on a light tan background.

Figure 21-10 shows a small ten-round box which has puzzled cartridge collectors for years. It is unquestionably "old," but is possibly the work of Bannerman or some other surplus dealer who repackaged the cartridges. Another possibility, given the short use of the Martin priming system in our service, and the inclusion of "U.S.A." on the label (meaningless for domestic consumption) is that it could date from the 1870–71 period, when we sold a great deal of arms and ammunition to France during the Franco-Prussian War.

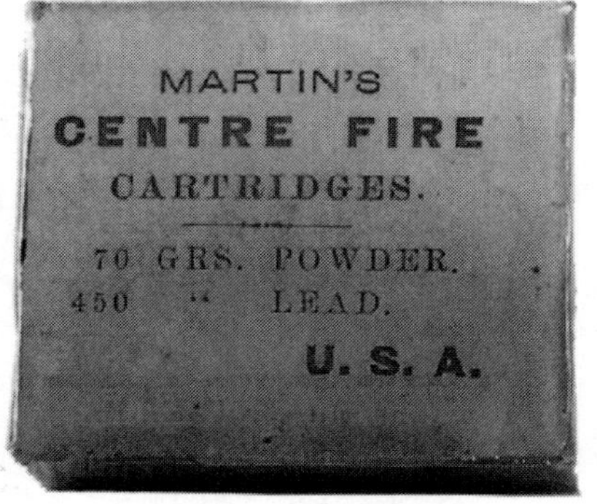

Fig. 21-10. A ten-round box of .50-70 Martin-primed cartridges that may have been repackaged for sale to France during the Franco-Prussian War, or else by a commercial surplus dealer of the late 19th century.

Another USCCo. contract box is clearly later (likely circa 1880) than the item shown in Figure 21-9, because it utilizes "Frazier's Patent Cartridge Pocket," patented on May 21, 1878, see Figure 21-11. This design uses a serpentine strip of rather heavy pasteboard (apparently wet-formed in a mold or die) in conjunction with a 3/8 inch x 1/4 inch wooden strip on one long side of the bottom of the box. The strip offsets the two rows of cartridges, making for easier withdrawal. This packing method was far superior to the more common interlocked (sometimes loose, sometimes glued) thin, notched pasteboard partitions.

Dimensions of the box are: 6-3/4 inches x 2-3/16 inches x 1-7/16 inches thick. The lid, which was a separate lift-off part, is missing. The label is printed in black ink on a light tan background. The twenty original cartridges have brass cases, a rounded rim, are not headstamped, and may be ether Farrington or Boxer primed; it is difficult to distinguish between the two. A very similar box is also known, but is more common, in .45-70. That box has the Frazier patent data on its bottom; the bottom of the .50-caliber box pictured here is blank.

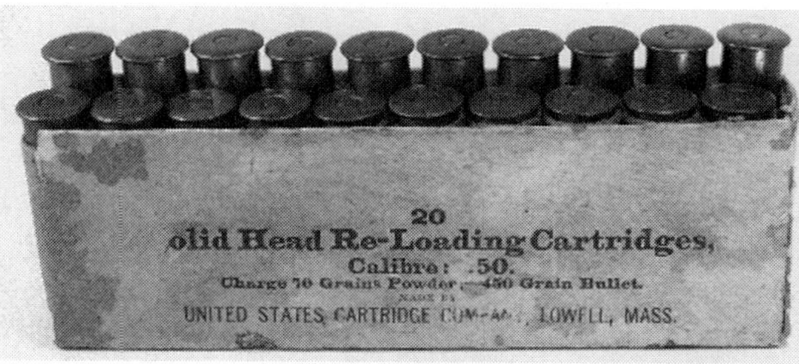

Fig. 21-11. USCCo. reloadable cartridges which used the "Frazier's Patent Cartridge Pocket." The separate lift-off lid to the box is missing.

.50-45-400 Cadet Cartridge

U.S. Model 1867 .50-45 Navy Cadet Rifle, Chapter 12

The short case was nominally 1.275 inches long (variations up to 1.332 inches are known), see Figure 21-12. It was originally developed for the U.S. Model 1867 .50-45 Navy Carbine (which was manufactured entirely by Remington) but was also used in the U. S. Model 1867 .50-45 Navy Cadet Rifle (Remington Patents). This arm alone, of the rifles discussed in this book, actually had a short chamber which would not accept *any* full-length .50 centerfire round, even those of reduced power. This would change in later carbines and cadet arms, as noted below.

Fig. 21-12. The .50-45 cartridges for the short-chambered 1867 Navy Cadet Rifle. Left, Martin-bar-primed with the 1.275-inch-long case; on the right, the Benet cup-primed cartridge with the 1.332-inch-long case.

.50-55-430 Carbine/Cadet Cartridge

U.S. Model 1866 (1867) Springfield Cadet Rifle, Chapter 5

U.S. Model 1869 Springfield Cadet Rifle, Chapter 7

Fig. 21-13. The .50-55-430 Carbine cartridge at left, was developed to moderate the recoil forces of the lighter-weight carbines issued to mounted troops from 1866–1872. It was also used in the U.S. Model 1866 (1867) and 1869 Cadet Rifles. The cartridge on the right is a .50-70-450 service rifle cartridge.

U.S. Model 1870 Springfield Carbine, Chapter 9

U.S. Model 1870 Remington Carbine, Chapter 14

U.S. Model 1870 Sharps Carbine, Chapter 17

U.S. Model 1871 Ward-Burton Carbine, Chapter 18

This cartridge saw service with the U.S. Model 1866 (1867) and 1869 .50-55 Springfield Cadet Rifles, as well as all four of the trial carbines, see Figure 21-13. The cartridge had a full-length (1.75-inch) case but was filled with only 55 grains of powder instead of 70 grains. The remaining space was filled with pasteboard wads. A lighter, blunt-nose 430-grain lead bullet was used instead of the 450-grain lead bullet of the service rifle cartridge. The shorter, blunter bullet made it possible to differentiate the carbine from the service cartridge as nearly all such ammunition from Frankford Arsenal was made without a headstamp.

Appendix A Known Serial Number Ranges .58- and .50-Caliber U.S. Springfield Rifles and Carbines		
Model	**Years Made**	**Serial Number**
U.S. Model 1865 .58 cal. "First Allin Alteration," both three- and two-band versions	1865	Not serially numbered
U.S. Model 1866 .50 cal. "Second Allin Alteration," both three- and two-band versions	1866	Not serially numbered
U.S. Model 1866 (1867) Cadet Rifle	1867	Not serially numbered
U.S. Model 1868 Rifle (1)	1868	1 - 125 (est.)
	1869	126 (est.) - 16000 (est.)
	1870	16001 (est.) to 52300+
U.S. Model 1869 Cadet Rifle	1871-1872	1-3422+
U.S. Model 1870 Rifle	1870-1873	Not serially numbered (2)
U.S. Model 1870 Carbine	1871-1872	Not serially numbered
U.S. Model 1865 Joslyn Rifle	1865 and 1871	M/1 - M/3025+ (3)
U.S. Model 1868 Remington Rifle, Transformed	1867-1868	1-504 Short (modified) version only
U.S. Model 1867 Remington Navy Cadet Rifle	1868	Not serially numbered
U.S. Model 1870 Remington Navy Rifle, both types	1870	Not serially numbered

The .58- and .50-Caliber Rifles

Model	Years Made	Serial Number
U.S. Model 1870 Remington Trials Rifle and Carbine	1871	Not serially numbered
U.S. Model 1871 Remington Army Rifle	1872	Not serially numbered
U.S. Model 1871 Infantry Rifle, converted Spencer	1871	Between 1 and 34500 (4)
U.S. Model 1870 Sharps Rifle and Carbine, Type I	1871	Between C30000 and C40000 (5)
U.S. Model 1870 Sharps Rifle (only), Type II	1871	1 - 300
U.S. Model 1871 Ward-Burton Rifle and Carbine	1871	Not serially numbered

1. This arm is unique as regards the dating of the breechblock with the actual year of manufacture, leading to an approximate division of the serial numbers. Numbers of the four carbines, and three pistols (if they have numbers) were not recorded.
2. See Chapter 8. The *vast majority are not* numbered; however, exceptions may exist.
3. See Chapter 10. The original 1865 rimfire guns should have all matching numbers; the 1871 centerfire conversions will have mixed numbers.
4. The new Springfield barrel was numbered to match the existing Spencer receiver from the range of Model 1865 carbines produced by the Burnside Rifle Co.
5. The new Springfield barrel was numbered to match the existing Sharps receiver from the range of "New Model 1863" production. The Sharps "C" prefix was not stamped on the new barrel.

Appendix B
Original Arsenal Finishes Applied to .58- and .50-Caliber Springfield Rifles, Carbines and Related Arms

Model	Receiver or Strap	Breech-block or Bolt	Hammer and/or Lock Plate	Barrel	Furni-ture	Stock
Model 1865 .58 cal. "First Allin Altera-tion," both three- and two-band versions	Black-ened (1)	Black-ened (2)	Color Case-Hardened	Bright	Bright	Oiled Black Walnut
Model 1866 .50 cal. "Second Allin Altera-tion," both three- and two-band versions	Black-ened (1)	Black-ened	Color Case-Hardened	Bright	Bright	Oiled (3) Black Walnut
Model 1866 (1867) Ca-det Rifle	Black-ened (1)	Black-ened	Color Case-Hardened	Bright	Bright	Oiled Black Walnut
Model 1868 Rifle	Black-ened	Black-ened	Color Case-Hardened	Bright	Bright	Oiled Black Walnut
Model 1869 Cadet Rifle	Black-ened	Black-ened	Color Case-Hardened	Bright	Bright	Oiled Black Walnut
Model 1870 Rifle	Black-ened	Black-ened	Color Case-Hardened	Bright	Bright	Oiled Black Walnut

The .58- and .50-Caliber Rifles

Model	Receiver or Strap	Breech-block or Bolt	Hammer and/or Lock-Plate	Barrel	Furni-ture	Stock
Model 1870 Carbine	Black-ened	Black-ened	Color Case-Hardened	Bright (4)	Bright	Oiled Black Walnut
Model 1865 Joslyn Rifle	Black-ened	Black-ened	Color Case-Hardened	Bright	Bright	Oiled Black Walnut
Model 1868 Remington Rifle, Trans-formed	Color Case-Hard-ened	Bright	Bright	Bright	Bright	Oiled Black Walnut
Model 1867 Remington Navy Cadet Rifle	Color Case-Hard-ened	Bright	Bright	Bright	Bright	Oiled Black Walnut
Model 1870 Remington Navy Rifle, TypeI/II	Color Case-Hard-ened	Bright	Bright	Bright/ Blued	Bright (5)	Oiled Black Walnut
Model 1870 Remington Trials Rifle and Carbine	Color Case-Hard-ened	Bright	Bright	Bright	Bright	Oiled Black Walnut
Model 1871 Remington Army Rifle	Color Case-Hard-ened	Bright	Bright	Bright	Bright	Oiled Black Walnut
Model 1871 Infantry Rifle, converted Spencer	Color Case-Hard-ened	Color Case-Hard-ened	Color Case-Hardened	Blued	Bright, Butt Plate color case-hard-ened	Oiled Black Walnut
Model 1870 Sharps Rifles (both types), and Carbine, Type I only	Color Case-Hard-ened	Black-ened or Color Case-Hard-ened	Color Case-Hardened	Bright	Bright	Oiled Black Walnut

Model	Receiver or Strap	Breech-block or Bolt	Hammer and/or Lock Plate	Barrel	Furni-ture	Stock
Model 1871 Ward-Burton Rifle	Black-ened	Bright	N/A	Bright	Bright	Oiled Black Walnut
Model 1871 Ward-Burton Carbine	Black-ened	Bright	N/A	Bright, some Blued	Bright	Oiled Black Walnut

1. This model did not have a receiver; the breechblock was attached to the barrel with a "strap" which was blackened.
2. A few Model 1865 breechblocks have been observed that were color case-hard-ened.
3. Issues of this model were directed to be made with "50% having oiled stocks, and 50% having varnished stocks."
4. An unknown, small number of barrels have been observed that were blued.
5. Some barrel bands were blued.

All but three of the rifles and most of the carbines discussed in this text had barrels and other parts finished in what was known as "National Armory Bright." This was a very fine "satin" finish, appearing almost "white" which was achieved by polishing the steel part with ever finer grits. It was *never* a "mirror" polish. Most such gun parts will now be found oxidized to a dullish, but not unpleasant, steely gray, and should be left that way. Do *not* overclean them, or attempt to repolish them, and never, ever, use Naval Jelly, or any other harsh chemical rust remover, on a collectible firearm!

Where browning (bluing) is mentioned, the author has tried to differentiate between the various processes used at the time, the two main ones of which were browning and heat bluing.

Browning (today called rust bluing) was achieved by degreasing the item, coating it with a mixture of nitric and hydrochloric acid and placing it in case or cabinet heated by steam pipes to eighty-five degrees Fahrenheit and in which sixty percent humidity was maintained. After rust formed on the part, it was rinsed in boiling water and then "carded" by it passing across a revolving bristle wheel to remove the flakes of rust. The formation of the oxide imparted a dark color to the metal. The process was repeated five times until the desired blue-black color was achieved.

Other steel parts such as external screws were often "blued" by being held in an open flame (heat bluing) until the part turned a dull red for a minute or slightly longer, then dipped in oil as a quench. The part so treated retained a bright, almost iridescent blue.

A variation on the heat bluing treatment was called "ash bluing" and was usually applied only to small steel parts. The part was coated with linseed oil and then dried after which hardwood ashes were sifted over it until the part was covered evenly. The coated part was placed in a furnace and heated until the color changed to blue. The part was removed, allowed to cool and the ashes brushed off. The part was then immersed in sperm oil and washed in a soda solution to remove the oil. The ashes, sticking to the linseed oil on the part as it cooled, prevented air from affecting the part and causing scale.

Many internal action parts, and their screws, were also "blackened" by dipping in oil which was then burned off in a flame. A dull black finish was imparted to the steel parts.

Another bluing process, called niter bluing, was used by commercial companies such as Sharps and Remington, but was not introduced at the Springfield Armory until about 1885. The parts subjected to niter bluing were usually small—barrel bands, screws, sears, etc. The part was dipped in a pot containing molten potassium nitrate (melting point, 607 degrees Fahrenheit) for several minutes, then withdrawn and dipped immediately into sperm oil, or other fine oil. The process was repeated until the desired color was reached and the part was then quenched in hot water. The process imparted a deep, iridescent blue color to the steel part. Niter bluing replaced ash bluing at the Springfield Armory.

Certain parts, such as the receiver and breechblock, were actually made of wrought iron and needed to be "case-hardened." Case-hardening was a process by which carbon was driven into the upper layers of the iron, or even low-carbon steel, to form a thin, high-carbon steel shell which resisted abrasion. The interior of the part remained iron and malleable to absorb shock. The technique used very high heat to drive carbon into the upper layers of the metal only.

The part to be case-hardened was cleaned and degreased and placed in an iron pot in the presence of animal charcoal (bone charcoal). The pot was heated to cherry red (around 1,100 degrees Fahrenheit). After a certain time at this temperature to allow the heat to "soak" all of the part, it was removed and quickly dipped into either an oil or water bath, or a combination of both. Oil imparted a blackish, sometimes mottled color to the steel. Water produced variegated colors of red and blue and yellow. Certain parts such as the receiver and breechblock were dipped in a combination oil and water bath. The oil, being lighter, floated on top of the water, and when the red-hot receiver passed through the oil, it was blackened. When it encountered the water, it was hardened.

For a more complete discussion of the manufacturing process, including finishes, the reader is referred to Ordnance Memoranda No. 22, *Fabrication of Small Arms*, originally published in 1878 by the U.S. Army Ordnance Department.

Appendix C

Some Principal Dimensions, etc., of the .58- and .50-Caliber Springfields and Related Arms (inches)

Model	Overall Length	Stock/ Forend or Forearm Length	Barrel Length	Muzzle Diameter (approx.)	Ramrod Length/ Retainer Style	Ramrod Head
Model 1865 .58 cal. (First Allin Alteration). Three-band version	56”	52-3/4”	40” (37-5/8” in bore)	0.775”	39-7/16” Swell	Tulip, flush
Model 1865 .58 cal. “First Allin Alteration,” two-band version	52”	48-3/4”	36” (33-5/8” in bore)	0.775”	35-7/16” Swell	Tulip, flush
Model 1866 .50 cal. “Second Allin Alteration,” three-band version	56”	52-3/4”	40” (36-5/8” in bore)	0.775”	38-3/4” Threaded	7-ring, cupped
Model 1866 .50 cal. “Second Allin Alteration,” two-band version	52”	48-3/4”	36” (32-5/8” in bore)	0.775”	34-3/4” Threaded	7-ring, cupped
Model 1866 (1867) Cadet Rifle	48”	44-3/4”	33” (29-5/8” in bore)	0.742”	32-9/16” Threaded	7-ring, cupped
Model 1868 Rifle	52”	48-3/4”	32-3/4” (32-5/8” in bore)	0.775”	35-5/8” Single Shoulder	7-ring, flush

Model	Overall Length	Stock/ Forend or Forearm Length	Barrel Length	Muzzle Diameter (approx.)	Ramrod Length/ Retainer Style	Ramrod Head
Model 1869 Cadet Rifle	49”	45-3/4”	(29- 5/8” in bore)	0.745”	32-5/8 Single Shoulder	7-ring, flush
Model 1870 Rifle	52”	48-3/4”	32 -5/8” in bore	0.775”	35-3/8 Single Shoulder Double Shoulder	7-ring, flush
Model 1870 Carbine	41-5/16”	29-3/4”	22”	0.789”	N/A	N/A
Model 1865 Joslyn Rifle, Both versions	52-1/8”	48-7/8”	35-9/16”	0.775”	35-15/16” (RF) 35-11/16” (CF)	7-ring, cupped
Model 1868 Remington Rifle, Transformed (original)	55-1/4”	Buttstock 13-3/8” Forend 34-11/16”	39-3/8”	0.775”	38-1/8” Swell	Tulip
Model 1868 Remington Rifle, Transformed (altered 1869)	51-7/8”	13-3/8” 34-5/16”	36”	0.775”	34-1/2” Single shoulder	7-ring, flush
Model 1867 Remington Navy Cadet Rifle	47-5/16”	12-51/4” 27-5/8”	32-9/16”	0.735”	31-9/16” Threaded (26 tpi)	7-ring, cupped
Model 1870 Remington Navy Rifle, both types	48-9/16”	13-1/2” 26-5/16”	32-5/8”	0.775”	29-1/2” Single Shoulder	7-ring, cupped
Model 1870 Remington Trials Rifle	52”	13-1/2” 31-5/16”	36”	0.775”	34-1/4” Single Shoulder	7-ring, flush

Model	Overall Length	Stock/ Forend or Forearm Length	Barrel Length	Muzzle Diameter (approx.)	Ramrod Length/ Retainer Style	Ramrod Head
Model 1870 Remington Trials Carbine	38"	13-1/2" 11-3/4"	22"	0.792"	N/A	N/A
Model 1871 Remington Army Rifle	52"	13-1/2" 31-5/16"	36"	0.775"	34-1/4" Double Shoulder	7-ring, flush
Model 1871 Infantry Rifle, converted Spencer	49-3/4"	14-1/16" 26-5/16"	32-5/8"	0.775"	29-3/8" Single Shoulder	7-ring, 1-1/2" short of muzzle
Model 1870 Sharps Rifle, Types I and II	52"	15-1/2" 31-1/16"	35-1/16"	0.775"	32-3/8" Single Shoulder	7-ring, flush
Model 1870 Sharps Carbine, Type I	38-15/16"	15-1/2" 11-3/4"	22"	0.792"	N/A	N/A
Model 1871 Ward-Burton Rifle	51-7/8"	48-3/4"	32-5/8"	0.775"	35-5/8" Single Shoulder	7-ring, flush
Model 1871 Ward-Burton Carbine	41-5/16"	29-3/4"	22"	0.785"	N/A	N/A

Appendix D

Glossary

Band Spring: An "L"-shaped piece of spring-tempered steel inletted into a groove cut in the forend and anchored with a pin (the short stroke of the "L") which held the barrel band in position against the forward thrust caused by recoil.

Bluing: the process of oxidizing, or rusting, a steel or iron part to prevent glare. Depending on the degree of polish, the color ranged from black to metallic blue. Contrary to popular opinion, bluing does not prevent further rusting. See also Browning.

Breechblock: The hinged top of the receiver. (Springfield used the two-word form at the time.)

Breechloading: A rifle or carbine in which a self-contained cartridge is inserted into the firing chamber or breech.

Bright (finish): Rifle barrels and other non-hardened steel parts before 1873 were polished to a dull natural metallic sheen at the Springfield Armory. The finish was termed "National Armory Bright."

Browning: Browning was the term for the bluing process used in the mid- to late 19th century.

Butt Plate: Metal cap or plate covering the butt end of the stock to prevent splintering.

Buttstock: Portion of the stock which fits against the shoulder.

Cannelures: Rings around the lower end of the ramrod to make it easier to grasp with the fingers.

Carbine: Correct term to use when referring to a short-barreled version of a rifle intended for mounted service.

Cartouche: A stamp impressed into wood or metal. The cartouche with the Master Armorer's initials, and after 1876, a year-date, indicated final approval of the arm by the Springfield Armory and acceptance by the U.S. Army's Ordnance Department.

Centerfire: A cartridge case containing a primer in the center of the base. When the primer is struck by the hammer, the flame produced ignites the main powder charge in the case.

Comb: Ridge along the top of the stock against which the cheek is placed.

Cone: 19th-century American term for the nipple in the breech of a percussion musket, rifle or pistol, on which the percussion cap was placed.

Forearm: Forward part of the carbine stock.

Forend: Forward part of the rifle stock.

Gun(s): In the U.S. Armed Forces, a gun refers to an artillery piece.

Heel: Top rear of buttstock.

In the White: See National Armory Bright; see also White.

Model 1863: Also called by some collectors, the Model 1863 Type I. This U.S. rifle musket is most identifiable by a date of 1863 or 1864 on the lock plate and round clamping barrel bands *without* barrel band springs.

Model 1864: Also called by some collectors, the Model 1863, Type II. The correct Ordnance Department model designation was "U.S. Model 1864." It is most identifiable by a date of 1864 or 1865 on the lock plate and rounded, non-clamping barrel bands held in place with barrel band springs.

Mortise: A hole or recess cut into wood or metal to receive a corresponding projection (tenon) to hold the two parts together.

Muzzleloader: A firearm which is charged with powder and ball through the muzzle.

National Armory Bright: A whitish, natural finish to steel or iron applied by the use of successive grinding grits to achieve a dull finish.

Nipple: Device threaded into the breech of a percussion musket, rifle or pistol on which the percussion cap was placed. Referred to as a "cone" in the 19th century.

Nose Cap: The metal tip or cap covering the end of the rifle stock.

Primer: An explosive compound contained in the base of the cartridge case which when struck by the hammer or firing pin, explodes and produces a very hot flame which ignites the main powder charge in the case.

Rear Sight: In most Springfield-produced long arms, the rear sight was mounted on the barrel, forward of the receiver. It had the following features: 1) aperture, the "V"- or "U"- shaped opening through which the shooter sighted at the target, 2) leaf or elevation leaf marked with range gradations, 3) leaf spring which held the leaf in position to which it was set by the shooter, 4) slide or elevation slide which moved along the leaf to set the range desired.

Rifle: A shoulder arm with a rifled barrel more than 24 inches long, firing a self-contained metallic cartridge and intended for use by foot soldiers.

Rifle Musket: Correct term for muzzleloading, percussion shoulder arms with barrels longer than 28 inches and shallow rifling. Rifle muskets in U.S. military service were manufactured at the Springfield Armory, or by contractors, between 1855 and 1864.

Rifling: Raised ridges and channels called "lands" and "grooves" inside the barrel, causing the bullet to spin about its axis, thus increasing its accuracy.

Rimfire: A cartridge whose priming compound is distributed around the circumference of the case at the base. The hammer is designed to strike the edge of the cartridge case to explode the primer.

Rolling Block: A patented breechblock system that used a pivoting, wedge-shaped block of steel to close the rear of the breech after a cartridge was inserted. When the hammer, located behind the breechblock, fell on the firing pin, it engaged a lip in the breechblock and held it closed against the rearward force of the fired cartridge. The shooter released the breechblock by pulling the hammer back.

Tip: Tip refers to the end of the forearm or forend. Also called a nose cap.

Thumbpiece: The lever used to unlatch and open the breechblock. (Springfield Armory hyphenated the term during the period.)

Toe: Bottom rear of the buttstock.

Trapdoor: A modern collector's term used to describe the breechblock used in the Allin system of firearms. The term comes from the fact that the cover-like breechblock had to be lifted up and forward to load a cartridge.

Trigger Guard: The protective strap of metal enclosing the trigger to prevent an accidental discharge.

White, or "In the white": Properly called National Armory Bright. The metal parts of the arm were polished to a dull finish and not blued, browned or darkened in any way.

APPENDIX E
BIBLIOGRAPHY

Ball, Robert W. D. *Springfield Armory, Shoulder Weapons, 1795–1968,* Antique Trader Books, Dubuque, IA 52004, 1997.

Benét, S.V., Bvt. Lt. Col. *Metallic Ammunition for the Springfield Breech-Loading Rifle-Musket*, Frankford Arsenal, Philadelphia, PA., 1868.

Butler, David F. *United States Firearms, The First Century, 1776–1875,* Winchester Press, New York, NY 10022, 1971.

De Haas, Frank. *Single Shot Rifles and Actions,* The Gun Digest Company, Chicago, IL 60624, 1969.

Dorsey, R. Stephen. *American Military Belts and Related Equipment,* Pioneer Press, Union City, TN, 1984.

Emerson, William H. *Encyclopedia of United States Army Insignia and Uniforms,* University of Oklahoma, Norman, OK, 1996.

Farrington, Dušan P. *Arming and Equipping the United States Cavalry 1865–1902,* Andrew Mowbray Publishers, Lincoln, RI, 2004.

Flayderman, Norm. *Flayderman's Guide to American Antique Firearms—8th Edition,* Krause Publications, Iola, WI 54990, 2001.

Frasca, Albert J., PhD, and Robert H. Hill. *1909 Catalog, Springfield Armory Museum, Arms & Accouterments,* Springfield Publishing Company, Carson City, NV, 1995.

Frasca, Albert J., PhD. *The .45-70 Springfield, Book II, Caliber .58, .50. .45 and .30 Breech Loaders in the U.S. Service, 1865–1893,* Frasca Publishing, Springfield, OH, 1997.

—. *Trapdoor Springfield Newsletter, Volumes 1–7,* Frasca Publishing, Springfield, OH, 1998–2005.

Fuller, Claud E. *Springfield Shoulder Arms, 1795–1865,* Francis Bannerman Sons / S & S Firearms, New York, 1969.

—. *The Breech-Loader in the Service, 1816–1917, Revised Edition,* N. Flayderman & Co., New Milford, CT, 1965.

—. *The Rifled Musket,* Bonanza Books, New York, 1958.

Gluckman, Colonel Arcadi. *U.S. Muskets, Rifles, and Carbines,* Otto Ulbrich Company, Inc., Buffalo, NY, 1948.

Hackley, F. W., W. H. Woodin, and E. L. Scranton. *History of Modern U.S. Military Small Arms Ammunition, Volume 1, 1880–1939*, Revised Edition, Thomas Publications, Gettysburg, PA 17325, and Robert T. Buttweiler, Ltd., Houston, TX, 1998.

Hardin, Albert N. Jr. *The American Bayonet 1776–1964,* Albert N. Hardin Jr., New Jersey, 1977.

Hicks, Major James E. *U.S. Military Firearms 1776–1956,* James E. Hicks & Son, La Canada, CA, 1962.

Hoyem, George A. *History and Development of Small Arms Ammunition, Volumes 1 and 2,* Armory Publications, Tacoma, WA 98444, 1981.

King, Moses. *King's Handbook of Springfield,* James D. Gill, Springfield, MA, 1884.

Logan, Herschel C. *Cartridges, a Pictorial Digest of Small Arms Ammunition,* Bonanza Books/Crown Publishers, Inc./The Stackpole Company, New York, 1959.

Mallory, Franklin B. *Serial Numbers of U.S. Martial Arms, Volumes*

1 through 4, Springfield Research Service, Silver Spring, MD 20914, 1983–1995.

Marcot, Roy M. *Spencer Repeating Firearms,* Northwood Heritage Press, Irvine, CA 92713, 1990.

McAulay, John D. *Rifles of the U.S. Army, 1861–1906,* Andrew Mowbray Publishers, Lincoln, RI 02865, 2003.

McChristian, Douglas C. *The U.S. Army in the West, 1870–1880, Uniforms, Weapons, and Equipment,* University of Oklahoma Press, Norman, OK, 1995.

Pitman, Brig. Gen. John. *The Pitman Notes on U.S. Martial Small Arms and Ammunition, 1776–1933, Volume 5, Miscellaneous Notes–Cals. .58, .50, .45, .30 . . .,* Paul E. Klatt/Thomas Publications, Gettysburg, PA 17325, 1992.

Poyer, Joe and Craig Riesch. *The .45-70 Springfield,* North Cape Publications, Inc., Tustin, CA 92781, 3rd Edition, 1999.

Reilly, Robert M. *American Socket Bayonets and Scabbards,* Andrew Mowbray Publishers, Lincoln, RI 02865, 1990.

Sawyer, Charles Winthrop. *Our Rifles,* Williams Book Store, Boston, MA, 1946.

Sellers, Frank. *Sharps Firearms,* Beinfeld Publishing, Inc., North Hollywood, CA, 1978.

Shaffer, James B., Lee A. Rutledge, R. Stephen Dorsey. *Gun Tools–Their History and Identification, Volumes One and Two,* Collectors' Library, Eugene, OR, 1992–1997.

Shields, Joseph W. *From Flintlock to M1,* Coward-McCann, Inc., New York, 1954.

Steffen, Randy. *The Horse Soldier, 1776–1943, Volume II, The Frontier, the Mexican War, the Civil War, the Indian Wars, 1851–1880,* University of Oklahoma Press, Norman, OK, 1978.

Treadwell, Major T. J. Ordnance Department, *Metallic Cartridges (Regulation and Experimental) as Manufactured and Tested at the Frankford Arsenal, Philadelphia, PA.,* Washington, Government Printing Office, 1873.

Waite, Malden D., and Bernard D. Ernst. *Trapdoor Springfield: The United States Springfield Single Shot Rifle, 1865–1893,* Beinfeld Publishing, North Hollywood, CA, 1980.

Wolf, J. Spencer, and Pat Wolf, *Loading Cartridges for the Original .45-70 Springfield Rifle and Carbine, Second Edition,* Pine Hill Press, Inc., Freeman, SD 57029, 1996.

Zupan, James. *Tools, Troopers & Targets,* J. M. Carroll & Company, Bryan, TX, and Mattituck, NY, 1985.

Reports and Manuals

Annual Reports of the Chief of Ordnance to the Secretary of War For the Fiscal Year Ended 1872 and 1875, Government Printing Office, Washington, D.C., as noted.

Cavalry Equipment 1874, Ordnance Memoranda No. 18, Government Printing Office, Washington, D.C., 1874.

Description and Rules for the Management and Cleaning of the Rifle Musket, Model 1863 for the Use of Soldiers, Government Printing Office, Washington, 1863.

Description and Rules for the Management of the Springfield Breech-Loading Rifle Musket, Model 1866, Springfield, United States Armory, 1867.

& Carbines of the Springfield Armory

Description and Rules for the Management of the Remington Navy Rifle, Model 1870, National Armory, Springfield, Mass., 1871.

Description and Rules for the Management of the Springfield, Remington, and Sharp's Rifle Muskets, Experimental Models, 1870, National Armory, Springfield, Mass., 1871.

Description and Rules for the Management of the Ward-Burton Rifle Musket, Model 1871, National Armory, Springfield, Mass., 1872.

Description and Rules for the Management of the Springfield Rifle, Carbine, and Army Revolvers, Calibre .45, National Armory, Springfield, Mass., 1874.

Fabrication of Small Arms, Ordnance Memoranda No. 22, Government Printing Office, Washington, D.C., 1878.

Infantry Equipment 1874, Ordnance Memoranda No. 19, Government Printing Office, Washington, D.C., 1875.

Manufactures at National Armory, 1872–1879, Government Printing Office, Washington, D.C., 1878.

Nomenclature Descriptive of the Rifle Musket, Model 1855, Reprint, Civil War Pictorials, Van Nuys, CA, 1966.

The Ordnance Manual For The Use Of The Officers of the United States Army, Third Edition, J.B. Lippincott & Co., Philadelphia, 1862.

Summary Statement of Ordnance and Ordnance Stores on Hand in the Cavalry Regiments in the Service of the United States (4th Quarter 1867, 4th Quarter 1870, 4th Quarter 1871, and 4th Quarter 1872).

About the Author

Richard A. (Dick) Hosmer has collected Springfield rifles of the period 1865–1915 since 1970, specializing in the "trapdoor" series of arms. A third-generation San Franciscan, he has lived in or near the San Francisco Bay Area since 1937, receiving his higher education at City College of San Francisco. His working life was spent in the construction industry, retiring in 2001 as senior project manager for a leading manufacturer of specialty doors.

He is a member of the Lincoln Rifle Club, a Life Member of the California Rifle and Pistol Association, and a Benefactor Member of the National Rifle Association.

He met Joe Poyer and Craig Riesch, the coauthors of *The .45-70 Springfield,* at the old Great Western gun show in Los Angeles. While doing some follow-up in the wake of their first volume in the ever-burgeoning "For Collectors Only®" series, he accepted Joe's challenge to write a book which would cover the Springfield arms of that period not covered in their work. The result is this book, *The .58- and .50-Caliber Rifles and Carbines of the Springfield Armory, 1865–1872*, which will be followed soon by *The Other U.S. .45-70 Caliber Springfield Rifles and Carbines*.

It is a source of pride to the author that his family was long associated with the Springfield Armory. According to a 1968 article in *Gun Report* written by Major Christopher Dvarecka, their last historian, there was at least one Hosmer on the Armory payroll from 1809 to 1915. That stands as the record for length of service by one family, at Springfield. Several of his nine cousins employed over these years were inspectors; one such cousin, George Hosmer, is believed to have inspected Ward-Burtons from the period of these books. Another relative, Frank L. Hosmer, inspected Government Model 1911 pistols during and after World War I for Colt's Patent Firearms Manufacturing Company.

In addition, the author was very pleased to have been instrumental in the return of a stolen U.S. Model 1873 Springfield Carbine to the Springfield Armory National Historical Site's museum in August 2005.

Books from

North Cape Publications®, Inc.

The books in the "For Collectors Only®" and "A Shooter's and Collector's Guide" series are designed to provide the firearms collector with an accurate record of the markings, dimensions and finish found on an original firearm as it was shipped from the factory. As changes to any and all parts are listed by serial number range, the collector can quickly assess not only whether or not the overall firearm is correct as issued, but whether or not each and every part is original for the period of the particular firearm's production. "For Collectors Only" and "A Shooter's and Collector's Guide" books make each collector and shooter an "expert."

For Collectors Only® Series

The .58- and .50-Caliber Rifles and Carbines of the Springfield Armory, 1865–1872, by Richard A. Hosmer ($19.95). In the immediate post–Civil War years, the U.S. Army and Navy were faced not only with the need to switch to breechloading rifles and carbines firing self-contained metallic cartridges, but to dispose of the suddenly obsolete muzzleloading, percussion rifle muskets and carbines in the most efficient and cost-effective manner possible. The Ordnance Department solved both problems by using the obsolete rifle muskets to build new cartridge-firing breechloaders.

But before the design of a "next-generation" rifle and carbine could be finalized, seven long years of research and testing, on the firing range and in the field, had to be undertaken by the military and civilian employees of the U.S. National Armory at Springfield, Mass. This book describes the .58- and .50-caliber rifles and carbines that were developed at the Springfield Armory between 1865 and 1872 and which led ultimately to the selection of the famed Allin "trapdoor" breechloading system. The Models of 1865, 1866, 1867, 1868, 1869, 1870, and 1871, both rifles and carbines, using the Allin breechloading system as well as the Joslyn, Remington, Sharps, Spencer and Ward-Burton that resulted, are described in detail along with their bayonets, ammunition, tools, and cartridge boxes.

Serbian and Yugoslav Mauser Rifles, by Branko Bogdanovic ($19.95). When Serbia won her independence from Turkey in 1878, the first responsibility of the new government of the small nation hemmed in by predatory powers on all sides was to develop a well-trained and disciplined military force and arm it with the best weapons then available. The Model 1871 Mauser, perhaps the best breechloading, bolt-action rifle available at the time, was the first choice. And so began a 125-year partnership with the Gebrueder Wilhelm and Paul Mauser Company that endured two world wars, two occupations and a Cold War, and that endures to this day.

It was Serbian gunsmiths who developed the famous “ring of steel” that enclosed the cartridge head and helped make the Model 1898 Mauser rifle a world standard in bolt-action, magazine rifles. As the years passed, Serbia/Yugoslavia adopted the latest Mauser models for their military forces until in 1956, the bolt-action rifle was discarded in favor of the semiautomatic, and then the automatic assault rifle. But the Mauser continues to be manufactured today as a military and police sniper rifle and as a sporting rifle for hunting and target shooting.

Thousands of Yugoslav Mauser rifles have been imported into North America in the last two decades; Mr. Bogdanovic’s book will help the collector and shooter determine which model he or she has, and its antecedents. Every Mauser that found its way into military service in Serbia/Yugoslavia is listed and described.

Swiss Magazine Loading Rifles, 1869 to 1958, by Joe Poyer ($19.95). The Swiss were the first to adopt a repeating rifle as general issue to all troops in 1869. The rifle was the Vetterli, a clever blend of Swiss and American engineering. In 1889, the Swiss adopted a small-bore rifle with a straight pull bolt and a box magazine, the Schmidt-Rubin, that somewhat resembled that developed around the same time for the British Lee-Enfield rifles. The design was so successful, that with relatively minor changes and upgrades, it remained in service until 1958 when it was replaced by a semiautomatic rifle. As with all the books in the “For Collectors Only®” series, there is a complete part-by-part description for both the Vetterli and Schmidt-Rubin rifles in all their variations by serial number range, plus a history of their development and use, their cleaning, maintenance and how to shoot them safely and accurately.

The American Krag Rifle and Carbine, by Joe Poyer, edited by Craig Riesch ($19.95). A new look on a part-by-part basis at the first magazine repeating service arm adopted for general service in American military history. It was the arm first adopted for smokeless powder, and it required new manufacturing techniques and processes to be developed for its production at Springfield Armory. The Krag was an outstanding weapon that helped define the course of American arms development over the next fifty years. In this new text, the Krag is redefined in terms of its development. Old shibboleths, mischaracterizations and misinterpretations are laid to rest and a true picture of this amazingly collectible rifle and carbine emerges. The author has also devised a monthly serial number chart from production, quarterly and annual reports from Springfield Armory and the Chief of Ordnance to the Secretary of War.

The Model 1903 Springfield Rifle and Its Variations (2nd edition, revised and expanded), by Joe Poyer ($22.95). Includes every model of the Model 1903 from the ramrod bayonet to the Model 1903A4 Sniper rifle. Every part description includes changes by serial number range, markings and finish. Every model is described and identified. Abundant color and black-and-white photos and line drawings of parts to show details precisely. 480 pages.

The .45-70 Springfield (4th edition, revised and expanded), by Joe Poyer and Craig Riesch ($19.95), covers the entire range of .45-caliber "trapdoor" Springfield arms, the gun that really won the West. "Virtually a mini-encyclopedia . . . this reference piece is a must," Phil Spangenberger, *Guns & Ammo*.

U.S. Winchester Trench and Riot Guns and Other U.S. Combat Shotguns (2nd edition, revised), by Joe Poyer ($16.95). Describes the elusive and little-known "Trench Shotgun" and all other combat shotguns used by U.S. military forces. "U.S. military Models 97 and 12 Trench and Riot Guns, their parts, markings [and] dimensions [are examined] in great detail . . . a basic source of information for collectors," C.R. Suydam, *Gun Report*.

The U.S. M1 Carbine: Wartime Production (4th edition revised), by Craig Riesch ($16.95), describes the four models of M1 Carbines from all ten manufacturers. Complete with codes for every part by serial number range. "The format makes it extremely easy to use. The book is a handy reference for beginning or experienced collectors," Bruce Canfield, Author of *The M1 Garand and the M1 Carbine*.

The M1 Garand, 1936 to 1957 (4th edition, revised and expanded), by Joe Poyer and Craig Riesch ($19.95). "The book covers such important identification factors as manufacturer's markings, proof marks, final acceptance cartouches stampings, heat treatment lot numbers . . . there are detailed breakdowns of . . . every part . . . in minute detail. This . . . volume is easy to read and full of identification tables, parts diagrams and other crucial graphics that aid in determining the originality of your M1 and/or its component parts," Phil Spangenberger, *Guns & Ammo*.

Winchester Lever Action Repeating Firearms, by Arthur Pirkle

Volume 1, The Models of 1866, 1873 & 1876 ($19.95)

Volume 2, The Models of 1886 and 1892 ($19.95)

Volume 3, The Models of 1894 and 1895 ($19.95)

These famous lever action repeaters are completely analyzed part-by-part by serial number range in this first new book on these fine weapons in twenty years. ". . . book is truly for the serious collector . . . Mr. Pirkle's scholarship is excellent and his presentation of the information . . . is to be commended," H.G.H., *Man at Arms*.

The SKS Carbine (3rd edition, revised and expanded), by Steve Kehaya and Joe Poyer ($16.95). *The SKS Carbine* "is profusely illustrated, articulately researched and covers all aspects of its development as well as . . . other combat guns used by the USSR and other Communist bloc nations. Each component . . . from stock to bayonet lug, or lack thereof, is covered along with maintenance procedures . . . because of Kehaya's and Poyer's book, I have become the leading expert in West Texas on [the SKS]," Glen Voorhees, Jr., *Gun Week*.

British Enfield Rifles, by Charles R. Stratton (each volume $16.95)

Volume 1, SMLE (No. 1) Mk I and Mk III (2nd edition)

"Stratton . . . does an admirable job of . . . making sense of . . . a seemingly hopeless array of marks and models and markings and apparently endless varieties of configurations and conversions . . . this is a book that any collector of SMLE rifles will want," Alan Petrillo, *The Enfield Collector's Digest*.

Volume 2, The Lee-Enfield No. 4 and No. 5 Rifles (2nd edition)

In Volume 2, "Skip" Stratton provides a concise but extremely thorough analysis of the famed British World War II rifle, the No. 4 Enfield, and the No. 5 Rifle, better known as the "Jungle Carbine." Includes all markings, codes, parts, manufacturers and history of development.

Volume 4, The Pattern 1914 and U.S. Model of 1917 Rifles

In Volume 4, the author describes the events that led to the development of the British Pattern 1914 Enfield and its twin, the U.S. Model of 1917 Enfield rifle. The M1917 was produced in and used on the Western front in far greater numbers than was the M1903 Springfield. Skip Stratton provides not only the usual part-by-part analysis of both rifles to show how the M1917 evolved from the Pattern 1914, but provides a cross-check of which parts are interchangeable. Included are the sniper and Pedersen Device variants.

The Mosin-Nagant Rifle (3rd revised and expanded edition), by Terence W. Lapin ($19.95). For some reason, in the more than 100 years that the Mosin-Nagant rifle has been in service around the world, not a single book has been written in English about this fine rifle. Now, just as interest in the Mosin-Nagant is exploding, Terence W. Lapin has written a comprehensive volume that covers all aspects and models from the Imperial Russian rifles to the Finnish, American, Polish, Chinese, Romanian and North Korean variations. His book has set a standard that future authors will find very difficult to best. Included are part-by-part descriptions of all makers, Russian, Chinese, American, Polish, Romanian, etc. Also includes all variants such as carbines and sniper rifles from all countries.

The Swedish Mauser Rifles (2nd edition revised), by Steve Kehaya and Joe Poyer ($19.95). The Swedish Mauser rifle is perhaps the finest of all military rifles manufactured in the late 19th and early 20th centuries. A complete history of the development and use of the Swedish Mauser rifles is provided as well as a part-by-part description of each component. All 24 models are described and a complete description of the sniper rifles and their telescopic sights is included. All markings, codes, regimental and other military markings are charted and explained. A thorough and concise explanation of the Swedish Mauser rifle, both civilian and military.

A Shooter's and Collector's Guide Series

The AK-47 and AK-74 Kalashnikov Rifles and Their Variations, by Joe Poyer ($22.95). The AK-47 and its small-caliber replacement, the AK-74, symbolize for Americans the now-defunct Soviet empire and its support for wars of "national liberation." Author Joe Poyer has examined and described the Kalashnikov rifle on a part-by-part basis, pointing out the differences between the various types of receivers and other parts, as well as the differences between the AK and AKM models. A detailed survey of all models of the Kalashnikov rifle from the AK-47 to the AK-108 is included as are descriptions of those Kalashnikov rifles manufactured by various countries from China to Switzerland. Accessories issued to the soldier from bayonets to web gear are included. Instructions on shooting, selecting telescopic sights, ammunition and troubleshooting round out the book.

The M16/AR15 Rifle (2nd edition, revised and expanded), by Joe Poyer ($19.95). The M16 has been in service longer than any other rifle in the history of the United States military. Its civilian counterpart, the AR15, has recently replaced the M14 as the national match service rifle. This 140-

page, profusely illustrated, large-format book examines the development, history, and current and future use of the M16/AR15. It describes in detail all civilian AR15 rifles from more than a dozen different manufacturers and takes the reader step-by-step through the process of accurizing the AR15 into an extremely accurate target rifle. Ammunition, both military and civilian, is discussed and detailed assembly/disassembly and troubleshooting instructions are included.

The M14-Type Rifle (2nd edition), by Joe Poyer ($14.95). A study of the U.S. Army's last and short-lived .30-caliber battle rifle which became a popular military sniper and civilian high-power match rifle. A detailed look at the National Match M14 rifle, the M21 sniper rifle and the currently available civilian semiautomatic match rifles, receivers, parts and accessories, including the Chinese M14s. A guide to custom-building a service-type rifle or a match-grade, precision rifle. Includes a list of manufacturers and parts suppliers, plus the BATFE regulations that allow a shooter to build a legal look-alike M14-type rifle.

The SAFN-49 Battle Rifle, by Joe Poyer ($14.95). The SAFN-49, the predecessor of the Free World's battle rifle, the FAL, has long been neglected by arms historians and writers, but not by collectors. Developed in the 1930s at the same time as the M1 Garand and the SVT38/40, the SAFN-49 did not reach production, because of the Nazi invasion of Belgium, until after World War II. This study of the SAFN-49 provides a part-by-part examination of the four calibers in which the rifle was made. Also, contains a thorough discussion of the SAFN-49 Sniper Rifle and its telescopic sights, plus maintenance, assembly/disassembly, accurizing, restoration and shooting. A new exploded view and section view are included. The rifle's development and military use are also explained in detail.

Collector's Guide to Military Uniforms

The "Collector's Guide to Military Uniforms" endeavors to do for the military uniform collector what the "For Collectors Only®" series does for the firearms collector. Books in this series are carefully researched using original sources; they are heavily illustrated with line drawings and photographs, both period and contemporary, to provide a clear picture of development and use. Where uniforms and accoutrements have been reproduced, comparisons between original and reproduction pieces are included so that the collector and historian can differentiate the two.

Campaign Clothing: Field Uniforms of the Indian War Army
Volume 1, 1866–1871 ($12.95)
Volume 2, 1872–1886 ($14.95)
Lee A. Rutledge has produced a unique perspective on the uniforms of the Army of the United States during the late Indian War period following the Civil War. He discusses what the soldier really wore when on campaign. No white hats and yellow bandanas here.

A Guide Book to U.S. Army Dress Helmets, 1872–1904, by Mark Kasal and Don Moore ($16.95).
From 1872 to 1904, the men and officers of the U.S. Army wore a fancy, plumed or spiked helmet on all dress occasions. As ubiquitous as they were in the late 19th century, they are extremely scarce today. Kasal and Moore have written a step-by-step, part-by-part analysis of both the Models 1872 and 1881 dress helmets and their history and use. Profusely illustrated with black-and-white and color photographs of actual helmets.

All of the above books can be obtained directly from **North Cape Publications®, Inc., P.O. Box 1027, Tustin, CA 92781** or by calling Toll Free 1-800 745-9714. Orders may also be placed by Fax (714 832-5302) or via e-mail to ncape@ix.netcom.com. CA residents add 7.75% sales tax. Current rates are any two books: Media Mail (7-21 days) $3.75, add $0.85 for each additional book. Priority Mail: 1-2 books $5.25 or 3-10 books, $8.95.

International Rates on request to E-mail address: ncape@ix.netcom.com.

Visit our Internet Website at http://www.northcapepubs.com. Our complete, up-to-date book list can always be found there.